CODING CARE

TOWARDS A TECHNOLOGY FOR NATURE

SABINE HIMMELSBACH
CHUS MARTÍNEZ (EDS.)

HEK (HOUSE OF ELECTRONIC ARTS)
INSTITUTE ART GENDER NATURE
HGK FHNW

HATJE
CANTZ

The magazine *Coding Care. Towards a Technology for Nature* comes with a history that is starting a little more than fifteen years ago.

In 2007, Karin Ohlenschläger, Yvonne Volkart, and Sabine Himmelsbach realized the exhibition *Ecomedia* at the Edith-Russ-House for Media Art, Oldenburg. The show traveled in 2008 to plug.in, Forum for New Media, Basel, the predecessor institution of HEK (House of Electronic Arts). *Ecomedia* was dedicated to the question of how artists use media technologies to gain a better understanding of our ecosystem. Analyzing and working with data, for example monitoring water quality, played a central role. While collecting and analyzing data was a new trend at the time, ten years later, when the three curators decided to do another exhibition on ecological change, data analysis has become an ubiquitous tool. The exhibition *Eco-Visionaries. Art, New Media, and Ecology After the Anthropocene* (2018–2020, at HEK and three other international venues) reflected what had changed since their last collaboration on *Ecomedia*. Politically—not so much, unfortunately. On the art side, however, there was an increased interest in the topic. The curators had chosen the provocative subtitle *Art After the Anthropocene* to assert, that we have already left behind the geological epoch marked by human activity and its profound impact on the environment and are looking at the world as a whole in a new way: at a superhuman world in which humans are no longer at the center. When working on the exhibition *Earthbound. In Dialogue with Nature,* curators Sabine Himmelsbach and Boris Magrini wanted to take up and continue these themes and debates. *Earthbound* has been commissioned by the Cultural Capital of Europe Esch 2022 in Luxembourg. It was first shown at the Möllerei in Esch-Belval and then traveled to HEK in Basel with a slightly adapted list of artists.

Many artists in the exhibition address the potential of media art as an extension of our scope of action, as relational models of communication with the superhuman world. Media technologies can become communication systems in which the relevant actors—from humans and animals to organic and inorganic substances—are in constant exchange and interaction. This can actually bring us closer together. Technologies help us for example to give voice to a river by monitoring pollution levels or tracking a flock of birds for a better understanding of their migratory habits.

For the show in Basel, a collaboration with the Institute of Art Gender Nature at the Academy of Art and Design FHNW, Basel was conceived to extend the context of the show by producing a magazine, focussing on developing a technology for nature, to use Coding as a means of Care. We also wanted to look towards regions, where there is a techno-scarcity and to allow for inventions, for fantasizing on the role technology could have in a particular culture and region. We decided to produce this publication as an extension of the experience provided by *Earthbound.* The exhibition has, indeed, already a catalog, then, why another publication?

Writing about an exhibition and writing around an exhibition are two complementary exercises. Since a while, we have been noticing how writers are considering a form of writing that is not a subject matter writing, such as analyzing a work or critically contextualizing it. We are witnessing a change in the tone and in the function of telling in order to engage with the energy flow created by artists interested in nature, leading to expand our perception and vision of potential futures, not only through art, but through writing: Telling as re-calling a memory; writing as a form of creating an assemblage and learning how to celebrate positive traits, such as joy; telling as a way to include us in the conversation with a scientist, with an expert that talks from a personal site; telling as witnessing the transformation of the field of research...

We are convinced that the directness of all the contributions do enhance a sense of inclusion and being-with the exhibition, even if one has not seen it. These texts are equivalent to an oral experience, they are meant to be read but, hopefully, they will give you a sense of being listening to friends, to individuals deeply committed to a transformation of the binaries that divide technology from nature, pragmatism from dreams, practicalities from values. There are no such separations, the world needs to come together and embody the rights of life in all its dimensions. *Coding Care* is located at the core of this wish.

We gathered texts by seven amazing authors. We would like to thank them for their inspiring contributions. A big thank you goes also to the artists of the *Earthbound* exhibition—it is their concepts, visions and ideas that inspire us. We are glad when they set us on our way with their projects and shake us awake to look for solutions.

We want to dedicate this magazine to our colleague and friend Karin Ohlenschläger, who passed away on August 24, 2022. The topic of art and ecology has been so crucial for her, and it feels like this current project is a continuation of the co-curated exhibitions, scientific exchange, and spirit we worked on together before. Exploring the dialogue between art, science, technology, and society and making it accessible to a wide audience was a big concern for Karin. We remember the intense and exciting discussions we had with her in the context of these projects, the passion with which she approached topics, developed relevant questions, and strongly advocated for an understanding of media art and its potential for society. We would have loved her contribution to this magazine and instead are honored to dedicate it to her.

Sabine Himmelsbach and Chus Martínez

EARTHBOUND – IN DI

BORIS MAGRINI

SABINE HIMMELSBACH

Questioning Our Coexistence with the Environment

We are confronted on an almost daily basis with dramatic images of extreme weather phenomena: floods, hurricanes, or forest fires due to extreme drought. Global warming is progressing, and it is factually proven that it is largely caused by humans, and that humans have become one of the most important factors influencing biological, geological, and atmospheric processes on the Earth. These are the facts we must confront ourselves with, when addressing the issue of environmental change from the perspective of contemporary art.

The exhibition *Earthbound*[1] at HEK (House of Electronic Arts), Basel features works by artists who, in dialogue with science, technology, and society, address the complex issues of our time and point to possible visions, opportunities, and solutions. Indeed, the urgent problems caused by human activity—global warming, pollution, the uncontrolled production of waste, and the loss of biodiversity and natural resources—require structural changes that are only possible through a radical rethinking of habits and mentalities. That a global capitalist system, geared towards the accumulation of wealth and unbridled production, is largely responsible for the above-mentioned calamities is a fact now widely acknowledged. As the historian and environmentalist Jason W. Moore states: "The rise of capitalism after 1450 marked a turning point in the history of humanity's relation with the rest of nature."[2] His analysis reveals that the logic of capitalism is premised on exploitation and appropriation, based on a dualistic idea in which nature is an entity extraneous to humankind, but also as a resource to be exploited and commodified. Today, this model is being discussed and called into question by theorists and artists who address ecological issues. They advocate a vision in which humanity and nature are not dissociated but rather considered to be an interconnected system. When, in 2018, Bruno Latour called for the need to establish a new ideological attractor that is based neither on a reactionary ecologism nor on a global liberalism,[3] we were in a context of a resurgence of the pre-pandemic ecological movement, led by the young and charismatic Greta Thunberg. At the same time, Donna Haraway asserted the need to create kinship with other species, defending the benefits of what she calls "speculative fabulation."[4] The concepts developed by Latour and Haraway have had a considerable impact on recent artistic production, particularly on artists dealing with ecological issues. The recent health crisis, the war in the Ukraine, and the advance of global warming dramatically highlight how global economic ties and dependencies have a decisive effect on both, bio-politics and the environment.

Some works in the exhibition deal with the human influence on the environment in concrete terms. In his video installation *Les Jardins cybernétiques* (2020), Donatien Aubert tells of the historical development of a reduction of natural phenomena to quantifiable models and the accompanying exploitation of natural resources. In an ironic gesture, his sculpture *Chrysalide N°3* (2020), takes up this reduc-

The recent health crisis, the war in the Ukraine, and the advance of global warming dramatically highlight how global economic ties and dependencies have a decisive effect on both, bio-politics and the environment.

tionist view of nature, which understands plants solely in terms of their utility value and is arranged in high-tech modules for domestic environments. The installation *The Intimate Earthquake Archive* (2016–2022), by Sissel Marie Tonn and Jonathan Chaim Reus lets visitors experience the strength of human-made earthquakes caused by gas wells in the Netherlands on their own bodies by means of specially designed vests. The earthquake data comes from archives and was made audible by the artists for physical perception. Our global ecosystem is under threat, and so are we humans. We need to find ways to be more sustainable in our use of resources. These changes require alternative models for thinking about our coexistence in relation to the environment. The transformations are exemplary of global developments that require reflection and adjustment regarding the balance between society, technological progress, and the environment.

Digital Archives and Techno-Utopian Models

The rise of digital technology, which affects all areas of life—from the economy to the private sphere and cultural production—cannot be seen as an abstract phenomenon that has no impact on the environment. It is necessary to continue to reflect on this process, on its meaning, on the possibilities it offers, and on the challenges it brings. Several works in the exhibition address the possibilities of current technological tools, such as the use of Artificial Intelligence (AI) to simulate possible future realities. With *Floralia I–IV* (2020), Sabrina Ratté presents digital memories of a plant life that may no longer exist in the future. In Alexandra Daisy Ginsberg's video installation *The Substitute* (2019), the white rhinoceros threatened with extinction is artificially brought to life by AI—in a virtual space that presents it to us without any natural context. The construction of perception by technological means in the digital age also plays a key role in the installation *Bark with a Trace* (2022), by Persijn Broersen and Margit Lukács: A digital piece of tree bark becomes itself a territory, a dematerialized landscape, thanks to the re-composition of thousands of high-resolution images. In Tega Brain's installation *Deep Swamp* (2018), AI is used experimentally and playfully to explore the problem of optimizing natural resources. Three AI-controlled 'software agents' optimize the aesthetic presentation of three swamp landscapes according to different objectives. In this work, it also becomes very clear that we humans have control over how technologies are used.

Embracing Complexity

Since the 1970s, a new way of thinking has emerged with regard to the interaction between humans and nature. The *Gaia* hypothesis, proposed by the biophysicist James Lovelock and the evolutionary biologist Lynn Margulis, describes the surface of the Earth as a complex, mutable system determined by living organisms. Instead of the "survival of the fittest," which goes back to the discoveries of Charles Darwin, the relationship between species comes to the fore. The concept of symbiosis, which sees and describes the Earth as a single organism, is becoming increasingly relevant. Lynn Margulis speaks of the "holobiont" and describes animals and plants no longer as autonomous species, but rather as a symbiotic network based on interactions between host and resident microorganisms living on and especially in it: "Life is an incredibly complex interdependence of matter and energy among millions of species beyond (and within) our own skin. These Earth aliens are our relatives, our ancestors, and part of us. They cycle our matter and bring us water and food. Without 'the other' we do not survive."[5] Artists are embracing these ideas of hybrid networks between humans and their environment, a future-oriented ecological principle of interconnectedness, a kinship of species.

Many works in the exhibition explore these complex connections with this "other," the more-than-human-world. The works of Erik

Bünger and melanie bonajo call for a better understanding of the animal kingdom and reveal the ignorance that often characterizes human behavior towards other species. In the video *Nature See You* (2022), Erik Bünger analyzes the video message of the gorilla Koko to the heads of state and government at the 2015 United Nations Climate Change Conference and shows how its gestures are interpreted in the sense of a transference to our human perspective. In contrast, melanie bonajo has children reflect on the respectful treatment of nature. She speaks of the lack of sensitivity that we need to regain in our dealings with nature. The audio-visual installation *Co_Sonic 1884 km²* (2021–2022), by Robertina Šebjanič immerses the viewer in the Ljubljanica River by adopting the point of view of the animals that inhabit it. By filming and recording the sounds of the river and the cohabitation of the different species, the artist sets out to offer a non-human point of view, while testifying to the impact on biodiversity caused by human activities related to the exploitation of the river. This sensitization is also emphasized by the Italian French philosopher Emanuele Coccia, who pleads for an ecological restructuring of society and stresses that humans must bid farewell to their special position within the ecosystem. In his words: "The world is the ever-changing result of the universal and cosmic intelligence and sensitivity of the infinite number of life forms." [6]

Transdisciplinary Communities and Plural Narratives

Several works in the exhibition seek to propose solutions to very specific problems. For example, the installation *KORALLYSIS* (2019–ongoing), by Gilberto Esparza consists of a modular structure made largely of ceramic that is designed to be placed on the ocean floor to facilitate the development of coral colonies, forming a symbiotic relationship with them. James Bridle has produced a series of five solar panels that have engravings representing five different species of radiolarians, single-celled marine organisms whose silica skeleton has a structure characterized by symmetrically ordered protuberances. In *Solar Panel 001-005* (2022), Bridle combines recent research into sustainable energy with shapes inspired by the more-than-human world to raise awareness of the need for collective, regenerative solutions to pressing ecological problems. Such projects are driven by a desire to address the question of the relationship between humanity and the environment, particularly in the light of the evolution of digital technologies and how these may or may not allow a new dialogue with the environment. They invite us to change our perception of being in the world and in coexistence with other species. It is no coincidence that these works are not of a traditional, purely contemplative nature, but are rather videos, interactive installations, Virtual Reality (VR) experiences, and sound sculptures, which are often the result of a research process carried out in collaboration with scientists and during long residency periods within the frameworks of "Art and Science" programs. In her analysis based on the sociological and economic study of the trade of matsutake mushrooms, the anthropologist Anna Lowenhaupt Tsching states: "New developments in ecology make it possible to think quite differently by introducing cross-species interactions and disturbance histories." [7] It is perhaps this plural narrative and a history of disturbances, rather than a linear and dogmatic one, that we also wish to tell with the exhibition *Earthbound*.

Artists are embracing these ideas of hybrid networks between humans and their environment, a future-oriented ecological principle of interconnectedness, a kinship of species.

Communicating with the More-Than-Human-World

The works in the exhibition are an invitation to explore a wide variety of possible coexistences between humans, the ecosystem, and autonomous technologies, and to understand humanity as a form of solidarity with others.[8] Many of the artists in the exhibition use the possibility of media technologies as extensions of our range of actions, as relational models of communication with the more-than-human-world, in which the relevant protagonists—from humans and animals to organic and inorganic substances—are in constant exchange and interaction. Technologies can actually bring us closer together. They can help us to give agency to a river by monitoring the level of pollution or by tracking a flock of birds and better understand their migration habits. Several works in the exhibition help to create awareness for the non-human-other. The physical experience and getting in touch are accordingly an important aspect of the exhibition—being involved, feeling with our own bodies.

Mélodie Mousset and Eduardo Fouilloux do this with their VR installation *The Jellyfish* (2020) by inviting the audience to interact with digital jellyfish. The virtual creatures first respond to their human counterparts by singing or humming and enter into a dialogue in which a synchronicity can be experienced on the level of sound. Reactions of humans and plants to music are juxtaposed in *Beyond Human Perception* (2020), by María Castellanos and Alberto Valverde. The work *Akousmaflore* (2007), by the artist duo Scenocosme (Grégory Lasserre and Anaïs met den Ancxt) invites the public to gently touch suspended plants, which respond with sounds of different intensities, showing how plants are sensitive to electrical flows and our presence. Marcus Maeder makes the earth itself tangible as a living organism with *Edaphon Braggio* (2019), a sound installation based on the fascinating sounds of soil animals in the Braggio region of Switzerland, which makes the ecosystem of this subterranean fauna tangible both, acoustically and haptically. The work *Input Field Reversal #2* (2022), by Ursula Endlicher translates HEK's surroundings into a digital 'browser-based' environment in which physical nature can be observed and manipulated. At the same time, the trees on site become an interface via Augmented Reality (AR) app: The images of the 'coded nature' of the digital plant variations can be accessed through tags on a selection of trees. A final example of using technology to better communicate and understand our environment is the work *Atmospheric Forest* (2020), by the artist duo Rasa Smite and Raitis Smits. In their VR installation, we become immersed in a forest and experience the emissions of trees, the typical smell of the forest, through the analysis of scientific data.

Bound to One Earth

The interactive works in the exhibition encourage the audience to take an active role, which at best contributes to an engaged reflection in the process of rethinking our relationship with nature. Regarding this empathy towards other species and the creation of hybrid networks between humans and their environment, there is also talk of "worlding with" or "becoming with," a future-oriented ecological principle of connectedness that goes back to the theories of Donna Haraway.[9]

The exhibition *Earthbound* strives to convey approaches to solutions. It is an invitation for participation and involvement. Nevertheless, art does not have the responsibility to save the world. Artists do not have to find solutions where politics cannot. At the same time, we are glad when they set us on our way with their projects and stir us awake to look for solutions. In all our interest, we have to strive for a better way of living together, a balanced coexistence between humans and the ecosystem, because a journey to other planets will not be an option (at least for our generation)—in the end, we are and will remain earthbound.

1
The exhibition *Earthbound – In Dialogue with Nature* has been commissioned by the European Capital of Culture Esch 2022 in Luxembourg. It has been shown at the Möllerei, Esch-Belval from June 4 to August 14, 2022 and afterwards travelled to HEK in Basel with a somewhat modified list of artists. Accompanying the exhibition, a catalog has been published with Hatje Cantz. This text is a slightly modified and shortened version from the catalog.

2
Jason W. Moore, *Capitalism in the Web of Life. Ecology and the Accumulation of Capital.* London 2015, p. 182.

3
Bruno Latour, *Down to Earth. Politics in the New Climatic Regime.* Cambridge 2018.

4
Donna Haraway, *Staying with the Trouble. Making Kin in the Chthulucene.* Durham 2016.

5
Lynn Margulis, *The Symbiotic Planet. A New Look at Evolution.* London 2013, p. 112.

6
Emanuele Coccia, *Metamorphosen. Das Leben hat viele Formen. Eine Philosophie der Verwandlung.* Munich 2021, quoted from: Leander Scholz, "Emanuele Coccia. 'Metamorphosen.' Die Ewigkeit der Körper," in: *Deutschlandfunk*, March 21, 2021. https://www.deutschlandfunk.de/emanuele-coccia-metamorphosen-die-ewigkeit-der-koerper-100.html (accessed on July 26, 2022). Translation from German by the authors.

7
Anna L. Tsing, *The Mushroom at the End of the World. On the Possibility of Life in Capitalist Ruins.* Princeton 2015, p. 5.

8
See also: Timothy Morton, *Being Ecological.* London 2018; James Bridle, *Ways of Being. Beyond Human Intelligence.* London 2022, p. 11.

9
Haraway 2016 (see note 4).

Installation view *Earthbound – In Dialogue with Nature*, 2022, HEK Basel. Photo: Franz Wamhof.

Donatien Aubert, *Chrysalide N°3*, 2020, *Les Jardins cybernétiques*, 2020. Installation view *Earthbound – In Dialogue with Nature*, 2022, HEK Basel. Photo: Franz Wamhof.

In order to make his structure plausible, he created of a
spatial allocation system which he called Flatwriter.

Persijn Broersen & Margit Lukács, *Bark with a Trace*, 2022. Installation view *Earthbound – In Dialogue with Nature*, 2022, HEK Basel. Photo: Franz Wamhof.

OF WOLVES AND GOATS AND SATELLITES

JAMES BRIDLE

On a wet day in June 1991, a few kilometers from Kananaskis Field Station in the foothills of the Canadian Rockies, a team of wildlife researchers captured and tagged a five-year-old female grey wolf. They named her Pluie, from the French word for rain.

Pluie was tagged with a new kind of device: a satellite collar, connected to a system called Argos, which was originally designed for tracking changes in the world's oceans. First launched in 1978, Argos collected temperature and pressure data from two hundred buoys drifting in the Antarctic Ocean. But the project had an unexpected side-effect: since the data also included the location of the buoys, marine and later terrestrial biologists realized they could use the same system for animal tracking. This was a quantum leap in capabilities. Previously, cumbersome VHF radio transmitters were used, which were only good for a few kilometers. Once researchers lost track of an animal, it might disappear from view forever.

This limit was a limit on our knowledge, and profoundly shaped the way we understood animal behavior. Previous VHF studies seemed to show that wolf ranges covered an area of a few hundred square kilometers, or at most a thousand in the wild territory surrounding national parks. In turn, this belief shaped ideas and policy around animal conservation programs, limiting them to the scale of national parks and the radius of VHF radio reception. When Pluie was connected to a global system, capable of tracking movement at the scale of a continent, a whole new understanding of animal behavior, and a whole new relationship with the natural world, became possible.

Argos is not very accurate—it can pinpoint an animal to around a mile—and it could be flakey, failing to report a viable signal for weeks at a time. When Pluie's signals started to come in, at first they seemed unbelievable, then jaw-dropping. For the first few months after her June capture, she stayed relatively close to the tagging site, but in autumn she suddenly upped and left the area, passing through Banff's parklands before turning west into British Colombia, and then south across the US border, into Glacier National Park. Passing east of the town of Browning, she entered the Great Plains. She travelled through the Bob Marshall Wilderness, a million acre roadless expanse of forests and mountains north of Missoula, before crossing the rest of Montana, and the northern part of Idaho, to reach Spokane Mountain in Washington State—some five hundred kilometers, as the crow flies, from Browning. A while later she headed north again, re-entering Canada from Idaho, somewhere near Bonner's Ferry. By December 1993, she had made it at least as far as Fernie, in British Columbia; a round trip covering more than a hundred thousand square kilometers.

The signal from Fernie was the last that Argos received from Pluie, and shortly after it was sent, the tracking team got a package in the mail containing the battery from Pluie's collar—with a bullet hole in it. They feared the worst, but Pluie was not dead, at least not yet: She turned up again, two years later and some two hundred kilometers away outside Ivermere, British Columbia, on the edge of Kootenay National Park, still identifiable by her now battery-less collar. This time, she did not survive: On December 18, 1995, she was shot and killed by a licensed hunter, along with an adult male and their three puppies.

In the six months that she was tracked on the move—and it is likely she undertook this journey multiple times over the course of her short life—Pluie had already traversed two nations, three US states, two Canadian provinces, and an estimated thirty different jurisdictions, from city limits to national parks, and Crown lands to First Nation territories. Her tracks crossed mountain ranges and forest wildernesses, as well as highways, golf courses, and private lands. She moved—apparently freely but often at risk from hunters, ranchers, and road traffic—across a previously unimaginable area. In doing so, she gave us new insights into the ways in which wild animals live their lives, amongst themselves, and amongst us. And she also showed us that the landscapes we imagine and set aside for animals are not remotely big enough.

One outcome of Pluie's incredible—but not unusual—journey was the development of a network of 'wildlife corridors' across the US and Canada, eventually connecting up to become one of the world's most ambitious environmental projects: the *Yukon to Yellowstone Conservation Initiative* (Y2Y). This project aims to bridge an expanse of some 3,200 kilometers, covering both national parks and the land inbetween them, via an interconnected system of wild lands and waters. The central idea of a wildlife corridor is to create a clear pathway, or network of pathways, along which animals and plants can move, unimpeded by human activity, in order to migrate, feed, and maintain diversity. Their creation and maintenance does require a change in the way we as humans plan, build, and live upon the earth.

Wildlife corridors can also be part of human healing processes.

Wildlife corridors can also be part of human healing processes. The largest nature reserve of Europe, and one of the longest wildlife corridors in the world, is the European Green Belt, a 7,000 kilometer network of parks and protected lands following the line of the Iron Curtain, which once separated Western Europe and the Soviet Bloc. The Green Belt was first proposed by German conservationists in December 1989, just a month after the Berlin Wall fell; today it stretches all the way from Finland to Greece. In places, old minefields still keep visitors on the paths, but the former 'death strip' is now a flourishing habitat and migration path for more than 600 species of rare and endangered birds, mammals, plants, and insects. One day the same might be true of the demilitarized zone (DMZ), between North and South Korea, a 250 kilometer long and four kilometer wide strip of land that has been virtually untouched by humans for more than six decades and is now home to millions of migratory birds and flourishing plant species, as well as endangered animals such as Siberian musk deer, cranes, vultures, Asiatic black bears, and a unique species of goats, the long-tailed goral. As in the thirty kilometer exclusion zone around the Chernobyl nuclear reactor, which scientists have called "an unintentional nature reserve," wildlife has flourished when humans have withdrawn. Despite high radiation levels in the zone, populations of elks, boars, foxes and deer are at least as high as in other preserves in Ukraine and Belarus—while one study suggests that wolves might be seven times more abundant.

Systems such as Argos, and newer technologies like GPS, allow us to see the world in an entirely new way, in scale and time. While traditional knowledge systems have always valued life and awareness in ways which balance human needs with our more-than-human kin, policy and management at the global scale at which humans now live, requires a macroscope: tools for seeing at the scale of the whole planet. These tools allow us both, to adjust our own lifestyle to better accommodate the flourishing of other species, and give us access to new kinds of knowledge we were not even aware of before.

In 2018 Russian cosmonauts mounted a three meter long antenna on the exterior of the International Space Station. This receiver was part of a new system called ICARUS, developed by researchers at the Max Planck Institute of Ornithology in Radolfzell, Germany, one of the country's network of advanced scientific research institutes. ICARUS takes advantage of new developments in sensor and broadcast technologies, allowing the most lightweight sensors yet constructed to be attached to all kinds of animals, and to track their movements across the planet. It has already given us a host of new insights into their behavior. The freely available app by ICARUS, "Animal Tracker," shows, in near real time, the locations of thousands of birds as they migrate and nest across Central Europe, and hundreds more across America, Africa, and Asia, as far north as Wrangel Island, off the coast of Siberia and as far south as New Zealand. It is thus providing invaluable data to biologists, as well as raising awareness of non-human lives among the wider public.

One experiment conducted by the ICARUS team gives some idea of what else we might discover. The researchers attached accelerometers to the collars of a number of different animals living in areas with high seismic activity: goats who roamed the slopes of Mount Etna in Sicily, and sheep, cows, and dogs around the city of L'Aquila in the center of Italy. These devices do not record locations or anything that humans typically refer to as 'personal data.' They solely transmitted the animals' degree of activity: how much they moved around, how placidly or excitedly they behaved.

When Mount Etna erupted at 10:20 pm on January 4, 2012, the team looked back over their data, and saw that the goats had become abnormally agitated six hours previously. Over the course of a two-year-study, they were able to predict another seven major eruptions. They found the same was true of earthquakes in L'Aquila: In the days and hours before earthquakes, the sheep, cows, and dogs would behave in unusual—but measurably unusual—ways. The more agitated they were, the closer they were to the looming epicenter. Together, they constituted an early warning system more powerful, more accurate, and more advanced than any other mechanism humans have devised. And these experiments were conducted with traditional radio tags: How much more powerful—and life-saving—such predictions will become now that the space-based antenna of ICARUS has come online, can only be guessed at.

There is no known mechanism for animal prediction of seismic activity—although it has been attested by folk wisdom for centuries. Perhaps, some scientists have suggested, their fur allows them to sense in some way the ionisation of the air caused by large rock pressures in earthquake zones. Perhaps, others argue, they can smell gases released from deep underground in the run-up to seismic events (the same mysterious gases have long been suggested as the mechanism which gave the Oracle at Delphi the gift of prophecy). We just do not know, but in some ways we do not need to. Rather than insisting on domineering and totalizing forms of knowledge which erase the lived experience of animals and other subjects of our experiments, we can instead use technology as a way of listening to them, and gaining from the knowledge they already possess.

Because of the variable ways that different animals react to natural phenomena, according to their size, speed, and species, the ICARUS team found it necessary to use correspondingly complex forms of analysis to pick up on the differences in the data generated from different tags at different times: a multitude of subtle and subtly variable signals. To do this, they turned to statistical models developed for financial econometrics: software designed to generate wealth by picking up on subtle signals in stock markets and investment patterns. I like to think of this as a kind of rehabilitation: penitent banking algorithms retiring from the city to start a new life in the countryside, and helping to remediate the Earth. From spy satellites to face-detection algorithms, the tools which currently conjure oppression and inequality can be turned around and put to beneficial uses. Despite the many problems complex technology causes for individual and collective agency, power, and privacy, we do not need to condemn everything we have learned about it in the search for justice and equality; we can repurpose it.

We also need to take care, to pay attention, and to act, together.

We also need to take care, to pay attention, and to act, together. One thing that became evident from these experiments was the need of many sources of data, many connected animals, to make accurate predictions. The behavior that suggested an oncoming tremor was not visible at the level of a single animal: it only became apparent when the data was aggregated. We are stronger, and more skilled and knowledgeable, when we act collectively, even if we do not know it ourselves.

One term which is often used to describe the use of large-scale tracking and communication of animal behavior is the 'Internet of Animals,' and while its potential to enlarge and shift our relationships with other species is vast, it comes with its own pitfalls and caveats. Just as the human internet has revealed the latent desires, foibles, weaknesses, and sheer strangeness of our own species, so the non-human internet will undoubtedly open us to encounters, experiences, and understandings we cannot presently conceive. It has the potential to revolutionize our awareness—but also to trap us into all-too-familiar matrices of oppression and domination. After all, it is not as if the 'Internet of People' has been an unalloyed success.

We will need to think carefully about the ways in which this new technology is deployed, used, and administered. In particular, we must not repeat the mistake of twentieth century technological determinism, which saw the role of high technology as producing the sole, unarguable answer to every problem. The use of trackers and other gadgets to amass vast amounts of data can sound suspiciously close to the kind of prediction and control that are characteristic of social and corporate media, and we will have to work particularly hard to ensure its applications are flexible, respectful, and appropriate. This endeavour will require the same kind of care, thoughtfulness, and creativity that marks artistic and activist interventions into technological debates today; a new and flourishing field of study which requires our urgent and direct attention.

Edited extract from James Bridle, *Ways of Being. Beyond Human Intelligence*. London 2022.

Erik Bünger, *Nature See You*, 2022. Installation view *Earthbound – In Dialogue with Nature*, 2022, HEK Basel. Photo: Franz Wamhof.

... Nature see you.

Erik Bünger, *Nature See You*, 2022. Video still. Photo courtesy of the artist.

... Koko cry.

Alexandra Daisy Ginsberg, *The Substitute*, 2019. Installation view *Earthbound – In Dialogue with Nature*, 2022, HEK Basel. Photo: Franz Wamhof.

Alexandra Daisy Ginsberg, *The Substitute*, 2019. Video still. Photo courtesy of the artist.

DARING TECHNOLOGY CODING CARE

CHUS MARTÍNEZ

The Computer-Dolphin

Even if we hold on for so long to the supremacy of the human species, some sort of a doubt hunted this believe: The question about animal's language and intelligence has been there since ancient times. The scholar Frans de Waal wrote a book that asked: *Are We Smart Enough to Know How Smart Animals Are?*[1] How do you ask another animal what it is thinking? Instinctively, humans not only develop tools to enhance their own capabilities, but we are a species equally obsessed with other species' intelligence. With a simple tool, the mirror, we proclaim that those capable of self-recognition possess a unique and distinctive type of intelligence. We first thought only humans belong to that category. But in the 1970s scientist Warden Gallup Jr. showed that our closest living relatives, the great apes, also shared this ability with us. "Oh, it's only the great apes that are going to show this because of our evolutionary connection." This mixed feeling of superiority and hope lasted for forty years until it was discovered that dolphins, too can recognize themselves in a mirror. Dolphins, contrary to apes, are not at all related to humans. They are mammals, but their body features and their life environment—they are aquatic—completely differ from ours. To discover that an animal has a sense of self forces us to inquire further about its evolutionary function. The emergence of self-awareness relates to other aspects of the cognitive abilities.

So, in dolphins like in humans, this ability starts emerging with their growing social awareness, their growing motoric skills, sensory-motoric development, their proprio-reception, which is the ability to track your body movements—you being aware of the movements of your body in a given space. And thus understanding that the image in the mirror is you, is directly connected with understanding what you are doing with your own body. Apparently, young dolphins show the emergence of self-directed behavior at the mirror and self-recognition even earlier than children—following an experiment conducted by Diana Reiss, Professor of Cognitive Psychology and Director of the Animal Behavior and Conservation Graduate Program at Hunter College, New York, and Rachel Morrison.[2] Children show it earlier than apes, but dolphins were showing it earlier than kids. Science growing interest in understanding the cognitive dimensions of these traits has had an enormous impact in our perception of the animal and also of nature in general. Knowing for sure that nature knows what we are doing with nature is making it unbearable to proceed with its abuse. It should not be the case, life is life and should not be destroyed, and even more though, since we are deeply affected by reciprocity and mutuality.

Another question is: How are these proliferating experiments and discoveries about the language and intelligence of animals affecting other realms of our impulse to define and design what intelligence is? If Life Sciences are strongly relating the perception of the body as a key element in sociability and self-perception, why is the development of Artificial Intelligence (AI) still so body-less? So social-less? What the mirror says is that we—humans, dolphins, and some fishes like the cleaner wrasse *(Labroides dimidiatus),* and certain birds—can recognize our self because we can recognize our bodies. And in doing so, we recognize the bodies of others. Recognizing the bodies of others is what allows us to live in society, in schools (fishes), in frocks (birds), etcetera. Why then did we take the body away and start to design a body-less intelligence that resembles the myth of a male chess genius or a mathematician, being just an organ, a brain and not a real body, a citizen, a friend or a lover.

The dreams of technology, besides a couple of butler-robots and sexy female disembodied voices in some movies, did remain quite outside the visions of transformation and the values of equality expressed in other realms of life. And that is probably why art and artists have become so important in the last two decades in inventing a new experience capable of overcoming the gigantic binary that still exists between technology-based and non-technology-based world-views. Indeed, in order to create a different AI, we need to give our AI the possibility of becoming a young dolphin being able to look at itself in a mirror.

Fantasy, Art, and Technology

I am always fascinated by how we fantasize: how fantasy allows the human to speculate different futures, but also reveals our desires to dominate. For example, we are keen on imagining how we might transfer certain traits—human intelligence and language—to programmed machines. But in the same dream, machines gain independence, take over, and misuse the power we pass on to them in order to destroy us. This fantasy rehearses our will to dominate and reveals our fear of another colonizer, one even more powerful and destructive than ourselves. AI is possessed by these fantasies. For this reason, I consider the exercise of producing different fantasies

Cecilia Bengolea, *Oneness*, 2019. Installation view der TANK, IAGN, Basel. Photo: Guadalupe Ruiz

a very necessary political activity. Fiction is different than fantasy and—more than ever—we need both.

The pandemic has revealed that disciplines based exclusively on fact and scientific data are coming, in their modern sense, to an end. Millions of people decided against the vaccine, for example, showing that trust cannot rely on scientific proofs only. Anxiety, and all the paranoias that grow in a hyper-sensitized social body, produces fantasies. The only way to counteract them is to channel them through other fantasies. I believe it is completely possible to imagine a technology and a science that are diverse and also attentive to how other species see and perceive. Imagine machines that could menstruate, or with sensorial devices mimicking the organs of animals rather than humans.

Fantasy is a technology that allows us to come closer to worlds we can easily visualize but cannot yet realize. Fiction allows for the transfer of emotion from the individual to the collective. If we combine these two technologies, we have a chance to create a world in which hybrid forms of organization can emerge. Right now, for example, we are still in a first phase, in which communities and groups claim—legitimately—their rights and their uniqueness, but still in order to defend themselves and their integrity. This defense is painful but necessary after centuries of abuse and oppression. But I am hopeful for a future in which we are able to form complex, solid, and empathic alliances between groups that may not share the same traits, experiences, or even species. Critical Theory or Political Philosophy helped us to do this in the past, and yet, even if I am a great lover of both, these disciplines, in their current forms, no longer serve the common good. That is why I am passionate about the role of art and artists. I see art as a substance that has, for centuries, been gathering the wisdom that we now need: an intelligence that is not programmed but shared.

The dreams of technology, besides a couple of butler-robots and sexy female disembodied voices in some movies, did remain quite outside the visions of transformation and the values of equality expressed in other realms of life.

Art knows that the force of co-creation, along with fantasy and fiction, allows us to experience freedom. Art interacts with inherited systems and their deep problems, and yet it also re-sensitizes materials. This is to move away from objects and towards life.

Lately, I have come to think that Western Art has—in this particular moment in time—used language as the 'big data' of art. Language as something to rely on, something to trust, something that conceptualizes life in a way that frees art from that burden. Yet art can do something as bold as placing the body at the core of its practice. How beautiful this is! What does it mean to have not only matter but bodies inside the practice of art? Science is motivated mainly by the human aspiration to be the only species to make decisions. We create instruments of observation and prediction, ostensibly for the sake of preservation, but in fact to enable us to control and make decisions on behalf of all other species and environments. By introducing the body, we introduce a co-creator. One that follows logics and paths that are different from the abstractions we superimpose on the world.

From cybernetics to performance, the body appeared to create a real fantasy: the body as a machine that cannot be entirely programmed by us. A body of cells connecting with other cells. All of a sudden art revealed that we never thought about the agency of other forms of life and we never designed systems for co-creating the future with them. This is just one example of how I think new epistemological parameters have been introduced through experience. Parameters that suggest, in a positive way, how we might come near to other things without destroying or extracting them. To say that nature should have autonomy might sound like a fantasy movie where the ocean talks and the clams dance. But this fantasy contains the key for a different form of governance, and a law-writing that imagines the rights of non-embodied entities.

I imagine art as an imaginative, evolutive substance capable of observing both, itself and life. An evolution that made art acquire morphic traits, enabling it, ahead of time, to take on the form our time requires. Like stem cells, art is able to join the organ that the body cur-

Cecilia Bengolea, *Danse au fond de la mer*, 2019. Performance, der TANK, IAGN Basel. Photo: Guadalupe Ruiz

rently needs. As in a bird flock, art is the intelligence that allows us to pirouette together in the skies, even if we cannot perform such moves on our own. Art activates the collective intelligence present in complex organizational systems. It is difficult to reduce this to exhibition-making or the market. Art should—and hopefully will—spread in a way that allows an absorption and understanding of the millions of transitions necessary to reach equality and freedom. This sounds idealistic: I would say it is a fantasy we need to live with joy. Joy is also an aspiration of art. I have the highest respect for joy.

Technology of Joy

One of the most amazing traits of art—and the virtue of artists—is giving space in love. Loving as the act of creating "a moving sea between the shores of your souls"[3] as poet Kahlil Gibran once wrote. These are hard times to be naïf, and yet the need to stay humble, sincere, and open are the conditions to safeguard freedom and possibility. We are constantly trapped in our paradoxical longing for intimacy and independence. This is a diamagnetic force—it pulls us toward togetherness and simultaneously repels us from it with a mighty magnet that can rupture a relationship and break a heart. That is why now, exactly now, we are exhausted, and a little bit scared, but mostly only anxious. It is crucial to give space in love, to be generous, to stay openhearted. It becomes an act of superhuman strength and self-transcendence. Can we do this? We cannot, or not alone. But art can.

If art can, technology can. And this is for me the important point to make. Art embracing love, embracing life is not the contrary or antagonistic to the industrialized machine created to enhance labor and consumerism. We need to invent a way of thinking those realms together. Technology could be both, a source of joy and a tool for life. Joy is a notion that affects all of life but has a philosophical and a cultural dimension that is quite unexplored. There have been certain moments in history—the late 19th century and the early 20th century—where the question of life, drive, passion, and politics has been crucial. We are in a moment that could be read as the opposite of this collective activation of forces and desires. Even in war, we seem unable to deactivate the impulse to center our feelings and attention upon ourselves. And yes, I want to talk about joy amidst a time where self-centering is an act of protection and a tactic to retain the freedom nullified by the mean circumstances we are experiencing. Because joy appears naive to many, I want to defend it as aspirational and anti-capitalist. In direct opposition to any totalitarian impulse, joy serves the task of reconnecting and reactivating a very damaged social tissue. All these artists commemorate and repeat—through their practice—the transformative act which is becoming more and more distant: the act of being there for one another. Joy declares the need to endure and enjoy the mutually exclusive traditions, trends, and individual positions. To find a way to de-radicalize, to find a path to say goodbye to violence, to exchange and transform is joy. There is nothing utopic about it. Joy is anchored to generosity. All the artists here are generous. That's it. That's what brings them together in my mind.

Excerpts from an interview published in summer 2022

Chus Martínez, "Towards Life," in: *Art Agenda*, New York 2022. www.art-agenda.com/criticism/478812/towards-life (accessed on September 23, 2022).

1

Frans de Waal, *Are We Smart Enough to Know How Smart Animals Are?* London/New York 2016.

2

Rachel Morrison, Diana Reiss, "Precocious Development of Self-Awareness in Dolphins," in: *PLOS ONE*, Sanya, China 2018. https://doi.org/10.1371/journal.pone.0189813 (accessed on September 23, 2022).

3

Kahlil Gibran, *On Marriage*, in: Kahlil Gibran, *The Prophet*. New York 1923.

María Castellanos & Alberto Valverde, *Beyond Human Perception*, 2020. Installation view. Photo: Sabine Himmelsbach.

PLANT'S STATE
Beyond Human Perception

María Castellanos and Alberto Valverde, *Beyond Human Perception*, 2020. Installation view *Earthbound – In Dialogue with Nature*, 2022, HEK Basel.
Photo: Franz Wamhof.

James Bridle, *Solar Panel 001 (Anthocyrtium hispidum)*, 2022, *Solar Panel 002 (Caclocyma petalospyris), 2022, Solar Panel 003 (Heliodiscus umbonatus)*, 2022. Installation view *Earthbound – In Dialogue with Nature*, 2022, HEK Basel. Photo: Franz Wamhof.

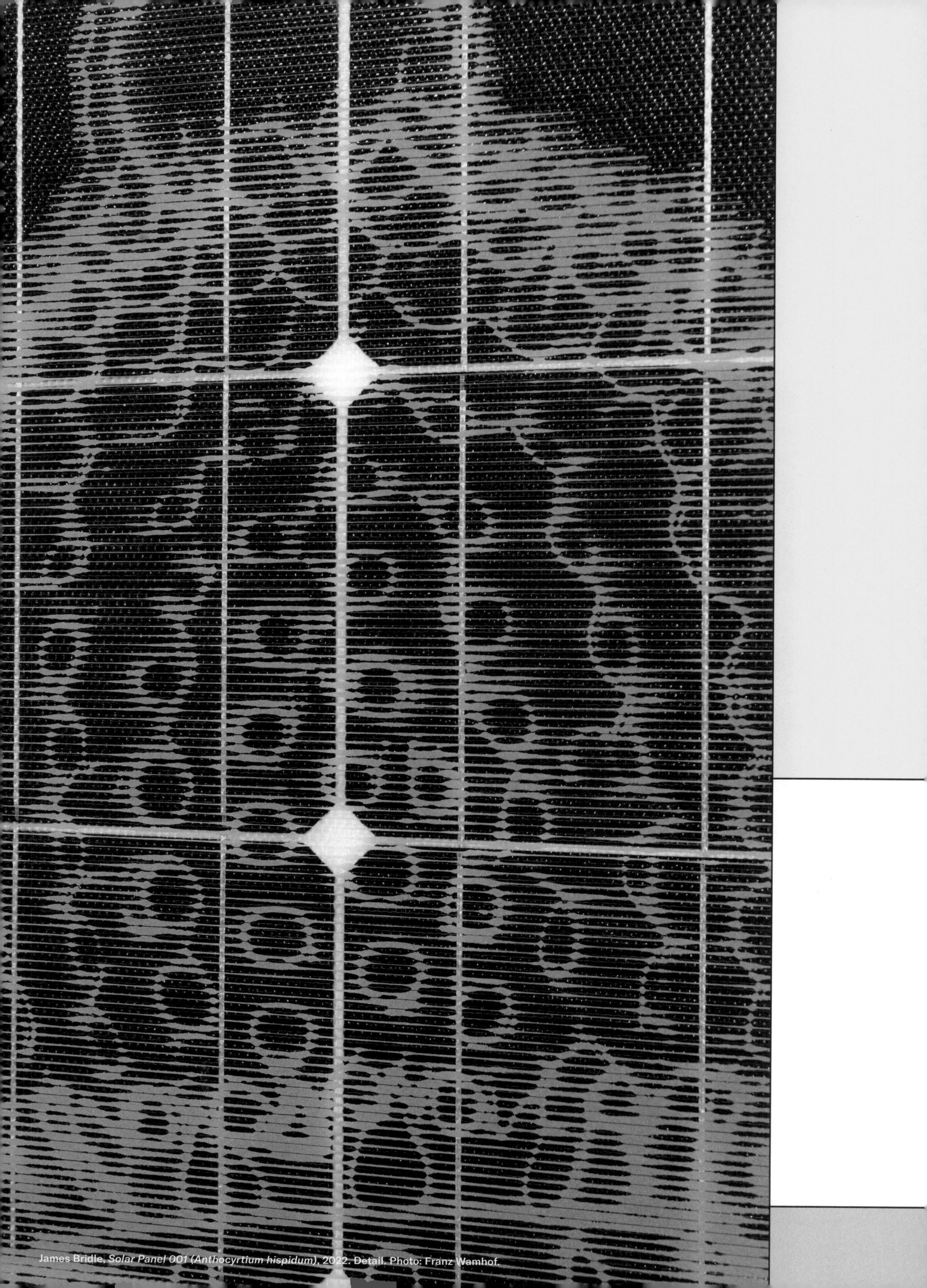

James Bridle, *Solar Panel 001 (Anthocyrtium hispidum)*, 2022. Detail. Photo: Franz Wamhof.

DEEP DREAMS, DEEP TIME

LUCIA PIETROIUSTI

In his 2016 book, *The Great Derangement*, novelist Amitav Ghosh meditates on the failures of fiction and imagination to tackle the reality of today's climate crisis. The book begins with the description of a familiar, yet uncanny, experience:

"Who can forget those moments when something that seems inanimate turns out to be vitally, even dangerously alive? As, for example, when an arabesque in the pattern of a carpet is revealed to be a dog's tail, which, if stepped upon, could lead to a nipped ankle? Or when we reach for an innocent looking vine and find it to be a worm or a snake? When a harmlessly drifting log turns out to be a crocodile?" [1]

Whether on its front line or sensing it at a distance, to encounter climate breakdown is, for Ghosh, to experience the sudden coming alive of earthly beings previously thought inanimate. The 'uncanny' is at work here: the familiar volcano that suddenly erupts, the earth under one's feet that trembles, the anomalous wave that sweeps away villages in a matter of minutes.

In "Haunted Geologies," an essay in Anna Tsing's and Elaine Gan's excellent anthology *Arts of Living on a Damaged Planet*, anthropologist Nils Bubandt articulates a similarly uncanny agency of the Earth itself, in the shape of a series of repeated eruptions, taking place since 2006, of a mud volcano in Java nicknamed Lusi by the local community. [2] The series of events have so far displaced some 40,000 people and released methane mixed with hydrogen sulfide and sulfur dioxide into the landscape, rendering the entire mud flat uninhabitable to all but the most extremophilic of bacteria.

Provoking a huge controversy, the eruption was attributed alternatively to natural or to anthropogenic causes: Did an earthquake cause the eruption, or was it rather the nearby activity of an oil-drilling company, that was to blame? Bubandt writes:

"Like Fukushima, Bhopal, Chernobyl, and other contemporary disasters where the forces of nature and human politics act to exacerbate each other, Lusi is the name for a monstrous geography haunted by the natural as well as the unnatural. But more so than other recent disasters with an anthropogenic component, the ontologies of the natural and the unnatural (whether human or spiritual kinds of un-nature) coalesce in Lusi's muddy ferment." [3]

In these instances, the uncanny slippage, far from a feature of the imagination alone, is precisely the irruption of the real as an animate thing, as a thing with agency: The Earth is not mute, and certainly not immutable. But in the case of Lusi, the mud volcano, the uncanny at play is a very specific, anthropocene uncanny, in which it is causality itself that is questioned. The disquieting slippage between the so-called 'natural' and the so-called 'human-made' blurs the lines of agency and animacy alike.

Is the uncanny 'natural', or is it 'human-made'? The question appears absurd in 2021, especially when viewed in the light of innumerable testimonies from representatives of Indigenous nations and communities around the world. In an entangled, continuous, vitalist paradigm, there is no opposition between these two poles, because, simply put, humans are entirely part of nature. Settler-colonial destructions of environments, lives, and livelihoods are all connected with one another, with the effect that the natural disasters that follow are both caused by human action, and they are the instantiations of a lively, animate Earth, speaking 'back'.

This reminds of Dipesh Chakrabarty's insight in his 2009 essay, "The Climate of History. Four Theses,"[4] and his subsequent publication, *The Climate of History in a Planetary Age.*[5] Chakrabarty argues, that the discipline of History, understood as contingent, human Historiography, needs to make space for the 'planetary' in the Anthropocene—whether it intends to or not. It is further inevitable, at this juncture in history, that we exist both in human time and in deep time—both in human space and in planetary and interplanetary ecologies, and all at the same time.

What could be the role of our technologies, then? Artificial Intelligence (AI) strikes us as both, uncanny, disquieting, and somewhat ridiculous when it breaks out of the conventions of human predictability and into what looks a lot like nonsense. Such is the case with GPT3, a new language-generating AI released in 2020. Powerful as it is, this new model is far more capable than more rudimentary chatbots at rendering language propositions that are uncannily close to human language.

What is at play here, when GPT3, like other chatbots, also begins to descend into senselessness or worse, hate speech, racist or sexist statements? Computer scientist Yejin Choi calls language-generating AIs "a mouth without a brain,"[6] that is to say, that algorithms can produce statements, but not understand the moral or ethical matrix within which these are being produced. Yet this view itself is also an exoneration of the anthropogenic nature of the issue: An algorithm trained on enormous amounts of data will respond and react to a set of human inputs. The algorithms that determine our interactions with social media will determine a path of communication that will generate extremes of languages, of positionalities, because their ultimate aim is not, as they would have it, to connect people, but to maximize their time of engagement.

Could we imagine an AI of repair, of forms of vitalisms that would bring us closer to our entanglements with more-than-human beings? This has been a question that has occupied the artist James Bridle and me in curating a section of the 2022 Helsinki Festival. We brought together artworks that think about the alternative possibilities of AI, not to alienate us from each other and from the complexity of the world around us, but rather to bring us closer to it. This is also a question that is at the core of Bridles recent book, *Ways of Being.*[7]

So is the uncanny 'natural,' or is it 'human-made'?

The possibilities of advanced technologies joining together in the effort of belonging to this Earth, rather than wishing us away from it, is a radical reframing. So far, this set of tools has been designed and bankrolled by corporate and ultimately nihilistic interests.

Artists, as the environmental engineer Tega Brain, are making this reframing tangible and visible in their work. The installation *Deep Swamp* (2018) consists of three tanks simulating wetland conditions, each modified and determined by three AI 'agents' with human names, whose encoded aims are manifestly different. According to the artist, Harrison aims for a natural looking wetland, Hans is trying to produce a work of art and Nicholas simply wants attention.[8]

In this way, Brain not only experiments with interactions between more-than-human agents (plant-based and artificial), she also poses crucial questions about how specific decisions in the design and training phases of algorithms lead to completely different behaviors and outcomes. Thus making it obvious,

Tega Brain, *Deep Swamp*, 2018. Installation view *Earthbound – In Dialogue with Nature*, 2022, HEK Basel. Photo: Franz Wamhof.

that intelligence and decision-making are entangled, and the notion of optimization, so far taken for granted in engineering, is, in fact a very contingent thing.

What is the impact created by the entanglement of dreams, imaginary worlds, empires, advanced technologies, corporate late capitalism, more-than-human beings across vegetal, mineral, and digital on the possibility of imagining, and maybe building, a 'mind' differently?

The thing that comes to mind here has to do with emergence. I have often wondered about this drive or involuntary tendency that we may have, to animate anthropomorphic things, while resisting the possibility of the consciousness or animacy to do so with non-anthropomorphic things.

In other words: Why are we so keen to assume that Alexa has a soul (try to ask: "Alexa, do you love me?" as my son did once, unprompted), but a tree does not?

Language, of course, is a clue. In the first edition of the festival and research project on more-than-human consciousness, *The Shape of a Circle in the Mind of a Fish* (2018), at Serpentine Gallery, London, Filipa Ramos and me focused on language and interspecies communication as one of the 'markers' of consciousness. It is thus no coincidence, that Alexa is designed to have a specific voice, to use human language and imitate human behavior. By contrast, the radical 'otherness' of vegetal communication places us in front of the unsettling experience of the untranslatable.

This is an issue that scientist Monica Gagliano, anthropologist Eduardo Kohn and political philosopher Michael Marder all address in their work in one form or another. Not so much: "How can we speak with more-than-humans in human-like language?" or: "How can we translate the language of a plant?" But rather: "Given a radical untranslatability with our more-than-human kin, can we come to an understanding of different ways of meaning-making, different forms of sense?" It is a perspectival approach, one that asks us to shift a little across the line into poetic or imaginary approximation, maybe even a little bit into dreams.

If meaning-making is at the core of these thinkers' work, relationality and interaction is its vehicle: Meaning, in a human as well as more-than-human context, does not articulate itself in a void, but in relation, in dynamic exchange, in co-evolved co-habitation, like flowers and pollinators, for instance.

Let's return to the question: why do we ensoul Alexa with so much greater ease than we can recognize the animacy, the meaning-making of a plant?

Another crucial issue is the idea of meaning itself being more complex than symbolic language, or even chemical or aural communication. Marder's astute poetic slippage helps us here: "If we were asked to provide a succinct definition of the language of plants, I would characterise it as 'an articulation without saying.' We would do well to recall that *articulation* has two, apparently unrelated, senses: expressing in words, and joining two or more things in space."[9] In articulating themselves through space and with one another, Marder argues, they create meaning, they make the worlds which they themselves inhabit.

Let us return to the question: Why do we ensoul Alexa with so much greater ease than we can recognize the animacy, the meaning-making of a plant? My experience of parenting has inflected my relationship to this question. When my child was born, I experienced very profoundly a dislocation of time, a de-scaling of time away from a single human life and towards some kind of ancestral and procestral, more-than-human time. I felt myself part of a chain of beings, one body born of another and still in some sense a part of it, as one single body growing out of itself. I also experienced my child as a unit, as a psychic whole—I had the experience of being able to recognize his entire being. It is only over time that I came to sense that the same must have been true of the experience my parents had of me—a whole. In contrast, I never conceived of my parents as psychic wholes, but only in relation to my own experience, my own memories.

I wonder then: Are we far more willing to recognize the psychic, ensouled wholeness of the beings we generate, precisely because we generated them, gave birth to them—when in contrast, we experience our ancestors only in fragments? Such is the continuum, the common denominator from tree to human to AI: that we are emergences. Meaning, language, and technologies are complex emergences from complex, more-than-human and ancestral worlds, and we perceive that complex world in fragments.

What kind of technology, then, can bring us to co-exist with complex and multiple, possible minds, in honor and recognition of that emergence? How do we all, with all our relations and in relation, to borrow both Winona Laduke's and Edouard Glissant's terms, begin to pick up the pieces?

1

Amitav Gosh, *The Great Derangement. Climate Change and the Unthinkable.* Chicago 2016.

2

Nils Bubandt, "Haunted Gelolgies. Spirits, Stones, and the Necropolitics of the Anthropocene," in: Anna Loewenhaupt Tsing, et al. (eds.), *Arts of Living on a Damaged Planet. Ghosts and Monsters of the Antropocene,* Minnesota 2017, pp. G121–G141.

3

Ibid, p. G124.

4

Dipesh Chakrabarty, "The Climate of History. Four Thesis," in: *Critical Inquiry,* vol. 35, no. 2, Chicago 2009, pp. 197–222.

5

Dipesh Chakrabarty, *The Climate of History in a Planetary Age.* Chicago 2021.

6

Matthew Hutson, "The Languague Machines," in: *Nature,* vol. 591, 2021, pp. 22–25, here p. 23. https://www.nature.com/articles/d41586-021-00530-0 (accessed on September 22, 2022).

7

James Bridle, *Ways of Being. Beyond Human Intelligence.* London 2022.

8

See the artis's project description, http://tegabrain.com/Deep-Swamp (accessed on September 21, 2022).

9

Michael Marder, "To Hear Plants Speak," in: Monica Gagliano, et al. (eds.), *The Language of Plants,* Minnesota 2017, p. 119.

Excerpt based on Lucia Pietroiusti, "Deep Time, Deepdream, Deepmind," in: *Woven in Vegetal Fabric. On Plant Becomings,* Luxembourg 2022, pp. 78–99.

Tega Brain, *Deep Swamp*, 2018. Installation view *Earthbound – In Dialogue with Nature*, 2022, HEK Basel. Photo: Franz Wamhof.

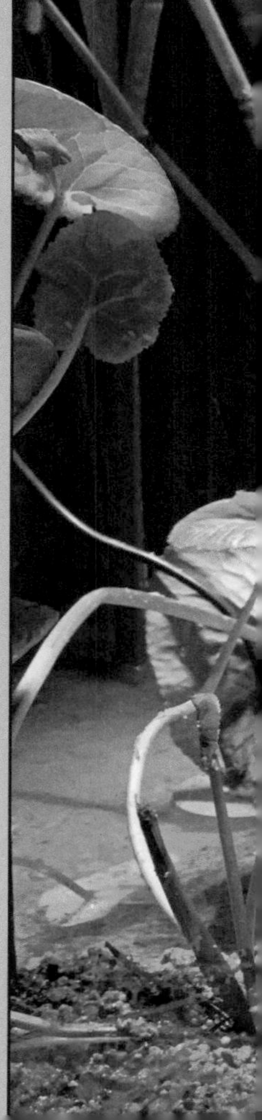

Tega Brain, *Deep Swamp*, 2018. Installation view. Photo: Sabine Himmelsbach.

Gilberto Esparza, *KORALLYSIS*, 2019–ongoing. Video still. Photo courtesy of the artist.
Gilberto Esparza, *KORALLYSIS*, 2019–ongoing. Installation view *Earthbound – In Dialogue with Nature*, 2022, HEK Basel. Photo: Franz Wamhof.

melanie bonajo, *Progress vs. Sunsets – Reformulating the Nature Documentary*, 2017. Video still. Photo courtesy of the artist.
melanie bonajo, *Progress vs. Sunsets – Reformulating the Nature Documentary*, 2017. Installation view *Earthbound – In Dialogue with Nature*, 2022, HEK Basel. Photo: Franz Wamhof.

Animals have a right to privacy, right?
-A lot of animals are filmed...

Scenocosme (Grégory Lasserre and Anaïs met den Ancxt), *Akousmaflore*, 2007. Installation view *Earthbound – In Dialogue with Nature*, 2022, HEK Basel. Photo: Franz Wamhof.

THE GROWING DOLOMITES

A CONVERSATION WITH HERWIG PRINOTH

FILIPA RAMOS

Inspired by the co-existence of people, animals, plants, minerals, stories, legends, languages and traditions in the Alpine region of Ladinia, the Biennale Gherdëina is a Contemporary Art Festival organized in the Val Gardena, in the heart of the Dolomites, in northeast Italy. Together with Lucia Pietroiusti, I co-curated its 8th edition, entitled *Persons Persones,* 2022. The title connects, through the same word, the global Esperanto of English and the rooted linguistic tradition of Ladin in paying tribute to the important efforts of recognition of non-human personhood in current environmental struggles. While looking at the legal rights of human and non-human persons, the 8th Biennale Gherdëina also found inspiration on the ancestral and contemporary movements of transhumance, in which humans, animals, and the microorganisms, seeds, and minerals they carry with them, shape and are shaped by the landscapes they traverse.

During the preparation of the exhibition, we had several encounters with experts from the most diverse areas, being introduced to the knowledge of regional linguists, artists, ornithologists, botanists, geologists, industrials, craftspeople, bakers, writers, and others. A person we kept wanting to spend time with was palaeontologist Herwig Prinoth. Even if we were fascinated, for instance, by the Dolomite legends told to us by writers and historians, and by the knowledge on migratory birds and travelling sheep shared by biologists and shepherding experts, Herwig's accounts captivated us from the beginning. He told us stories of vanished nautiloids, forever sleeping cave lions, and the endless adventures of the young Dolomite mountains across the ages.

With the accuracy of a scientist, the enthusiasm of a nature lover, and the generosity of a storyteller, Herwig devoted much of his time to us, sharing his passion for these mountains with the artists, curators and the team of the Biennale. The facts and narratives he transmitted permeate the whole show, as they shaped our lines of investigation and became artworks. His knowledge and the delicate balance with which he combined science and narrative give a very good example of how the temporalities of science, the precision of technological knowledge devices, and the fascinating lines of story-telling make the world a place understood and experienced differently. The following dialogue with him attests and celebrates such triangulation at its best.

Filipa Ramos: I would like to begin with a question about the marine origin of the Dolomites, a fact that amazed myself and Lucia Pietroiusti, and made us radically change the way we perceive and understand the Alpine landscape around us. Can you tell us about the aquatic past of the mountains that characterize the landscape of the Val Gardena?

Herwig Prinoth: The Dolomites did not come into being within the area where they are today. About 250 million years ago, they were situated at least 4,000 kilometers further south, at the heart of the Tropical Zone, near the Equator. From that era until 20–30 million years ago, the area of the Dolomites was almost always submerged beneath the sea. It was a warm tropical sea called Tethys, a nymph in Greek mythology. As a result, apart from the *Quarzporphyr* of the Resciesa High Alps and the Val Gardena sandstone, virtually all the rock strata around Ortisei were formed in the sea. This is why almost all fossils in the Dolomites are of marine origin. The animals whose skeletons (calcareous sponges, bacteria and, in the more recent parts, corals) built the cliffs of the Dolomites, which resembled modern-day coral reefs, could only live and grow in the sea.

If we look at mountains such as the Sassolungo, the Sella, the Cir and the Odle, we see not only an Alpine landscape of rare beauty but also a seabed with enormous cliffs fossilized about 240 million years ago. To see similar sights elsewhere, it would be necessary to travel in a submarine around the Caribbean or the Seychelles, where there are huge atolls divided by underwater valleys. Just like in the Dolomites, only here the sea has disappeared.

FR: What you are telling sounds as much as science as poetry or science fiction. It's this capacity of yours to recount scientific facts as if they were stories and return time to a human scale that fascinated us. You've made us understand how incredible Geological History is. In this balance between exact knowledge, scientific research and your awareness of a story whose main characters are the mountains, we ask ourselves how a palaeontologist like yourself relates to the huge difference in scale that exists between in-depth analysis and study for the writing of papers, and knowledge of the incommensurable spatial and temporal extension of the subjects you address.

HP: To tell you the truth, it's virtually impossible to understand and come to terms with the vast time gap that separates us from the events I study during my palaeontological research trips. I have calculated that if every year represents a teaspoon of water, the 251.9 million years that separate us from the End-Permian Extinction would be 3,778.5 cubic meters of water, which corresponds to a swimming pool of 40 by 30 meters, filled with water 3.14 meters deep. Alternatively, if a grain of rice represents a year, we would have 16.37 tonnes of rice. It's a terribly long period of time, but it becomes manageable if we look at pictures showing the various strata and their ages. This certainly helps to reason in tens of millions or even hundreds of millions of years. Though it has to be said that grasping geological periods as long as these is extremely difficult, nonetheless. Maybe we were able to get our heads around them better when we used the old Italian lire currency. Then we were used to thinking in millions, but now that we have the euro, the trick doesn't work anymore.

FR: At this point, storytelling helps us to imagine such abstract numbers and a scale that is far superior to the human one. I imagine that, for a scientist like you, these narrative elements emerge in encounters with specific objects—a plant or animal fossil, for example—that come to the surface from a distant past and, as forms, become present and concrete. Is the knowledge of an Alpine palaeontologist based a lot on these forms of fossilized life found in the mountains or are they just details of a scientific narrative that begins with the more abstract systems, proportions and formations of which you spoke earlier?

HP: Fossils underpin all palaeontological research, providing us with a lot of information, for example, on how the environment must have been in given periods of geological history. But a fossil isn't enough on its own: It's necessary to delve further and see how this organism interacted with the environment and, above all, with the other animals with which it coexisted. Only fossil associations can give us satisfactory answers. If, for example, all the fossil bivalves in a stratum are of the same species and are typical of a shallow sea with a higher-than-normal salinity level, we can be certain that the environment was hypersaline and subject to stress, with only minimal biodiversity. If instead, in a stratum, we find many bivalves of different species associated with other animals such as brachiopods and nautiloids, in this case we can be sure that the environment was healthy with a rich biodiversity. It's necessary therefore to consider both, the single fossil and fossil associations. Data on fossil associations may then of course be analyzed with a variety of statistical methods to draw broader conclusions about the climate, the environment, the formation and the extinction of a species.

FR: Around Ortisei—an area that we are particularly interested in since it's where the Biennale Gherdëina takes place—is it possible to find the fossils of solitary and gregarious creatures? What can we learn from the distribution of fossils in these parts?

HP: The Ortisei area is rich in fossils. Here are situated the ridges that stretch from the Val d'Anna to the top of Monte Balest and the one that climbs from Val Cuecenes to the top of Monte Seceda. These rocky projections hide strata of Val Gardena sandstone containing carbonized plants, with remains of the *Ortiseia*—a conifer—and footprints of reptiles in particular, while the *Bellerophon* Formation contains many marine fossils, including those of nautiloids, relatives of the *Nautilus*, a living fossil, and the gastropod known as *Bellerophon*. The Werfen Formation contains many fossil bivalves, among which diverse species of *Claraia*. In the Buchenstein Formation, it is possible to find ammonites and an exceptional discovery, part of the skeleton of an *Ichthyosaurus*: a marine reptile similar to a dolphin. Nowadays, organisms usually form communities and groups, and it is rare to find solitary ones. And the same must have been true hundreds of millions of years ago, insofar as it is usual to find many fossils of the same species. If a given fossil is very rare, it usually means that the creature lived in an unsuitable environment, while in other environments the same fossil may be present in abundance. It should be added that marine molluscs are very numerous because their shells fossilized easily in the marine environment, whereas the fossils of vertebrates (reptiles, fish, amphibians) are rare because it's much less likely for the skeleton of a vertebrate to be preserved. From the distribution of fossils in strata, we can reconstruct the habitat in which an organism lived. Every environment and every type of sediment contain particular fossils. If, for example, we find traces of a reptile inside a stratum, we can be sure that the stratum itself was formed in a terrestrial and not in a marine environment. And if we find marine fossils, the environment was certainly not terrestrial. If we find fossilized plants together with shells, we can be sure that islands must have existed somewhere from which plants were eradicated by tropical storms and deposited on the seabed. But a single fossil tells us little, and it is always necessary to view all the fossils contained in a stratum to be able to reach certain conclusions.

We are the first species to be conscious of its own existence and for the first time in the history of the Earth—thanks to our intelligence—we can counter the danger of extinction and find remedies.

FR: Nautiloids fossilized in the rocks at the highest altitudes in the Dolomites. Even though we are talking about events with a completely different timescale, bearing in mind the present climate situation and the drought that an area like the North of Italy is experiencing, do you believe that these proto-animals offer us an advance warning of what the future holds for the world?

HP: Almost all the animals I study and find fossilized in the rocks of the Dolomites are now extinct. Only very few organisms like the *Nautilus* and a bivalve known as *Acharax* have remained nearly unchanged over the last 252 million years and since then they have managed to survive all natural disasters.

Unfortunately, we are facing a devastating climate crisis as well, if we fail to find remedies for it. But we aren't like the poor old dinosaurs who had to suffer the impact of meteorites passively without being able to do anything to save themselves. We are conscious of the danger, we know we are responsible for the greenhouse effect with our emissions, caused by transport, industry, agriculture and so on. Since we are conscious of the danger, we can do something about it. We are the first species to be conscious of its own existence and for the first time in the history of the Earth—thanks to our intelligence—we can counter the danger of extinction and find remedies. If instead we do nothing against the greenhouse effect, I'm afraid we won't be as lucky as the *Nautilus* and the *Acharax*. Probably we, too will meet the same end as the vast majority of fossils we find in the rocks of the Dolomites: We'll become extinct.

FR: Your previous description of fossil groups in the Dolomite area is the story of a veritable pre-historic ecology, showing how the relationship between organisms and their habitat is fundamental in explaining why specimens may be numerous or only isolated. At the same time, and making a leap forward of probably several millennia—which seems enormous for us, but isn't for pre-history—we were fascinated by the story you told us about the cave lion bones found in a cave in the Dolomites. But in all these transformations, in which some marine animals ventured onto *terra firma,* adapting to a new life that demanded a different anatomy, when was it that the mountains assumed the form we know today? And is this form stable or is it constantly changing?

In nature, everything is evolving: Nothing stays still.

HP: In nature, everything is evolving: Nothing stays still, everything is transformed. We often don't even notice it, because these transformations only take place over a very long period of time. Since a human life lasts little more than a hundred years at most, we don't notice changes that often only become clear after thousands, even millions of years. The Dolomites are growing in height by about 0.75 millimiters a year. Erosion manages to compensate for the growth and so the true height of the mountains does not change that much. But if we want to go up to Conturines cave at an altitude of 2,750 meters, after ten years, we have to ascend an extra 7.5 millimeters, without of course realizing it. But after 10,000 years we'd have to go up an extra 7.5 meters. After a million years it will have increased by about 750 meters—quite a slog! The Dolomites assumed their current appearance somewhere in the region of 15,000 years ago, at the end of the last Ice Age. At the climax of the last Ice Age, about 22,000 years ago, the ice was about 1,500 meters thick. Only the highest Dolomite peaks peeped out of the ice. Given that ice never stands still and runs along valleys, and the mixture of rock and ice at the base of a glacier is incredibly abrasive, the Dolomites continued to be unearthed and shaped by the pressure of the ice for thousands of years. The Dolomite landscape was shaped by ice and, as today, it is nothing other than what remained after the glaciers had dismantled most of the dolomitic rocks. The rocks that remain today are the most ancient, while the younger ones were completely eroded and dislodged. The snow, rain, and wind of the last 10,000 years have only touched-up the Dolomite landscape slightly.

Sabrina Ratté, *Floralia I–IV*, 2020. Video still. Photo courtesy of the artist.

Sabrina Ratté, *Floralia I–IV*, 2020. Installation view *Earthbound – In Dialogue with Nature*, 2022, HEK Basel. Photo: Franz Wamhof.

4

Ursula Endlicher, *Input Field Reversal #2*, 2022. Installation view *Earthbound – In Dialogue with Nature*, 2022, HEK Basel. Photo: Franz Wamhof.

BODY WITH MORE ORGANS

INGO NIERMANN

To act on behalf of nature, our empathy cannot be trusted. Rather than feeling with non-humans, we have to feel as them by receiving their nerve and brain impulses and transforming them into our own experiences. The 'Internet of Beings' will allow us to overcome the liberal 'body without organs,' equipped with a variety of exchangeable protheses, and turn into a body with additional non-human organs.

In recent years it has become common practice to grant nature basic rights. In 2008, the constitution of Ecuador included the rights of nature to exist, persist, maintain, and regenerate its vital cycles.[1] Bolivia changed its constitution accordingly in 2009. The first case of two humans acting as legal guardians of a natural entity—the Vilcabamba River in Ecuador—was successfully argued and won in 2011. Since then, numerous countries have passed similar treaties, laws, and judicial decisions. The United Nations have been discussing the rights of nature in a series of "Harmony with Nature" dialogues by the General Assembly, and have started an initiative for a *Universal Declaration of the Rights of Nature.*

All these basic rights do not care about individual, mortal lives, but about whole habitats and species that can potentially thrive forever. As we can not take care of all creatures on Earth—or even just manage to leave them alone—the whole ecological movement has been following this approach.

Soon, one will look back at our time as the last years of an unbelievable ignorance, not just about human-inflicted harm against nature, but within nature, too. To promote empathy with our environment, we shield ourselves from its cruelty—even though the evidence of it is accumulating exponentially. In the wake of the miniaturization of information technology in the 1960s, biologists started to equip animals with small radio tags to track their movements. Later came cameras, GPS, and electronic sensors. As a result of those observations we know, for instance, that cats eat only thirty percent of their brutally killed prey. Shouldn't we breed cats that are less bloodthirsty? And do the same with wild animals, as a result of similar findings?

Breeding, mass surveillance, genetic engineering—humans have always been keen to apply new technologies to non-humans rather than to themselves. Soon it will be possible to automatically equip all kinds of creatures—down to microbes—with information technology, process the input, and control behavior accordingly. We—or rather our machines—will not just be able to listen and talk to these creatures, but to read and re-write their minds and nervous systems as well. With the 'Internet of Things' comes the 'Internet of Beings.' The question will not be whether to alter non-humans as well or not, but how to prevent the algorithms from acting in an all-too-patronizing manner.

The realm of information technology is guided by the ineradicable belief that with continuously growing streams of information we will finally be able to reach a god-like level of certainty about our world. Mechanism design will guide us to the best-case-scenario for everybody. It is a belief that goes back to the ideas of Vladimir Vernadsky and Teilhard de Chardin of the 'noosphere' in the 1920s, rooted in Plotinus's prophecy of the world soul, itself rooted in Platonic idealism. All three share the separation of a pure sphere of reason from an impure world of matter. This already went wrong once on a large scale with the ordinary internet. Exponential computational powers go hand in hand with exponential means of manipulation.

When the Digital Minister of Taiwan, Audrey Tang, asks for "a form of democracy that includes rivers, mountains, animals and other nonhumans, giving these agents a political voice via an avatar,"[2] it will not be enough, as she optimistically proposes, for people to "start to devise their own games that are win-win for everyone involved. If they detect a win-lose dynamic, people will simply turn it into a game, working together to develop a better mechanism instead of focusing their energy on making other people lose so that they can win."[3] Why should they? As long as some enjoy it when others lose, or prefer not just to win, but to win ever more. One would have to exchange these people with ideal game participants. And who would be in charge of such a transformation?

To avoid ever-greater concentrations of power, we have to create a democracy that allows non-humans to actually have a saying. Prior to the 2012 edition of Documenta, its artistic director, Carolyn Christov-Bakargiev asked: "What would a world be if the bees were voting, and how could the bees vote? Or a strawberry, how can a strawberry vote?"[4] When I asked her how this could work, her ideas were following a similar approach as with rights of nature: humans would act as their proxies.

On which basis would these people judge, for instance, if a strawberry prefers to be eaten or rot in the bush? Probably the strawberry does not know itself, and is unable to choose between different options (not just in the sense of flourishing or not, but in having an understanding of itself). So how could we make the strawberry eligible, without basically replacing it by something completely new—maybe equipped with wheels or an exoskeleton, as imagined by cognitive scientist Joscha Bach in his article "How to Build an Artificially Intelligent Plant" (2016)?[5]

No false modesty here: Whatever we turn natural entities into, they will be our fabrication, just as we already domesticated and cultivated and transformed most natural entities that surround us. Rather than declaring these things independent beings, it would be more 'natural' to acknowledge them as part of us. Instead of us belonging to our environment as something bigger than us, or our environment belonging to us as property—we would be something bigger than us. The liberal deterritorialization as an imagined 'body without organs' with an endless variety of exchangeable protheses would be followed by a reterritorialization of the body with a potentially endless number of extra organs.[6] Empathy is a limited mechanism. We can misinterpret others or completely ignore them, but when it comes to our own organs, their signals shoot straight to our brain. We do not feel empathy with our organs, we just feel them.

No false modesty here: Whatever we turn natural entities into, they will be our fabrication, just as we already domesticated and cultivated and transformed most natural entities that surround us.

The 'Internet of Beings' will allow us to send data generated from entities all around the world right to our limbic system. We could decide by ourselves which ones become our additional organs, and how their signals transform into feelings. These organs could be part of our own legal property, or someone else's. Everybody would have the right to feel everything.

Some might prefer masochistic scenarios that give them pleasure whenever their entities of choice are harmed. Some might find relief in self-destruction. But most would likely go for entities, they are able to identify with, and that they want to be all right. Most might identify with obvious attractions (such as dolphins or trees), some might challenge themselves and go for entities that used to make them feel uncomfortable (like jellyfish or mould). Our choices and settings would be full of anthropocentric biases. Our environment would still be misrepresented. But at least we would get in line with our needs and wishes for it.

1

Republic of Ecuador, Constitution of 2008, chapter 8. https://pdba.georgetown.edu/Constitutions/Ecuador/english08.html (accessed on September 19, 2022).

2

Joanna Pope and Audrey Tang, "After Zero-Sum. Serenity for Democracy," in: *Arts of the Working Class,* Berlin 2020, no.120. http://artsoftheworkingclass.org/text/after-zero-sum-serenity-for-democracy (accessed on September 19, 2022).

3

Ibid.

4

From a conversation at Creative Time Summit. *Confronting Inequity,* October 12, 2012, New York. https://creativetime.org/summit/2012/10/12/carolyn-christov-bakargiev (accessed on September 19, 2022).

5

http://bach.ai/how-to-build-an-ai-plant (accessed on September 16, 2022).

6

The terms body without organs, deterritorialization and reterritorialization refer to concepts developed by philosophers Gilles Deleuze and Felix Guattari in their work *Capitalisme et Schizophrénie,* Paris 1972, vol. 1 and Paris 1980, vol. 2.

“YOU CAN'T ALWAYS SEE THE WIND BUILDING THE LANDSCAPE”* BUT IT DOES BUILD IT

YINA JIMÉNEZ SURIEL

The reflections below are a compilation of my thoughts on a friendly technology programmed by other living beings. In a particular way, it has assisted the human species in its adventures to build different worlds. These thoughts are excerpts of three texts published in June 2020,[1] August 2021,[2] and September 2022.[3] The reoccurence of this technology in my writing is no coincidence, since it appears to be a backbone of my curatorial research: *La Historia de las montañas*. My research evolves around two axes of work: emancipation and how the human species builds imagination. I am understanding both as symbolic and material structures, allowing the management of the life of living beings.

In the context of this magazine, I find it particularly interesting to contribute with this compilation of thoughts about certain aesthetic tools developed by human communities in different time periods and contexts. They follow the purpose of expanding the perceptual system of the human species, seeking to make it more opaque, more dense and multidimensional by creating other imaginative structures of how to live in relationship.

Desde La Tierra de altas montañas

The imagination in which we live has dried out. Although, we still inhabit its ghost, the world in which stability and the binary were the foundation of everything, died a long time ago. The current crises and socio-political scenarios on the planet only evidence its death.

With this awareness, we must conquer the crisis and create new visions, but this time recognizing movement as an intrinsic characteristic in the life of every living being. And we do not have to go very far: The ocean is the space from which we can create that moment radically different from the present. Take any human idea and put it in the ocean; it will surely end up collapsed or diluted, because that which is rigid, has no place in water.

fo fi fu fá fo fi fu fuán fo fi fu fuán fo fi fu fuán fo fi fu fuán fo fi fu fuán fo

Undoubtedly, we are talking about an emancipatory process, or rather the continuation of what has been going on over the last three centuries. Are not the American and African independences and the creation of their own National States the result of the search for a movement that denies colonialism?

In my part of the world, this search for movement began to be thought of in the Maroon communities in the high mountains, consisting of escaped enslaved people. So, paradoxical as it may sound, in the Caribbean, continuing emancipation across the ocean cannot be separated from the history of the mountains.

The first thing the mountains tell us, is that the key tools to disarticulate old ideas and create new ones are not supernatural devices, but sonorities and active listening. Used as a wind instrument, the *fututo,* made from the shell of a snail, made the formation of communities of resistance possible in the Caribbean where Indigenous and African populations fled from enslavement. Tracing the origins of regional independence processes of the 19th and 20th centuries leads us to the *fututo* as an articulator, with different ways of sounding but always calling for freedom.

19°27'26.13'N 70°41'19.68'W

The *fututo* is an artifact producing sonorities. It is a wind instrument whose origin is the sea. The Caribbean Sea is both passageway and portal. A passageway is a path leading from the underground to resistance. The portal, on the other hand, is a magical device that allows for escape, that allows for travel. Thus, both terms, passageway, and portal, operate in resistance and alliance. In the *fututo,* resistance and alliance have their own sonic forms.[4]

Resistance sounds like:

fufí fí fí fufifi fufifi fí fufifi fii fufifi fufúuuuu fofuuuuuuu fouuuuuufofuuuu fooofufooofofó fofufofó fofofó fofouu fofofó fofuuu fofofuuuu fufúuuuuu

Alliance sounds like:

fo fi fu fá fo fi fu fuán fo fi fu fuán fo fi fu fuán fo fi fu fuán fo fi fu fuán fo fi fufú fuán fo fo fuán fofofofó fuán fo fofofó fuán fofofofó fuán fueeeefúan fueeee fuánfueeefuán fueefuán fufufufú fuán fuefufufú fuán fue tufofó fuán fue fufufofó fuán fueee fuán fuee fuán fuee fuán fueee fuán fueee fuán fuefueee fuán fufuee fuán fufufo fuán fufufofó fuán fufufo fuán fufufofo fuán fofufufo fuán

The underwater mountain range called Sierra de Bahoruco in the Dominican Republic, the Massif de la Hotte, and the Massif de la Selle in Haiti know this soundtrack very well. Many of the escape trails on the island lead to its jungle. For a long time, the call to resistance was heard from there. And it still resonates—surely it resonates—and could be amplified by our own bodies if we inhabit active listening.

The *fututo* is an artifact for the Maroon experience that has been used by both, original and enslaved, African communities, followed by those of us who survived them. Its sounds traveled throughout the region due to the circularity of the Caribbean at that time. Today, few know how to touch the *fututo,* but from it and its records, we are building other artifacts for resistance and alliance, which are imminent to create other possible fictions.

Walking back and forth, and becoming aware of these Maroon processes, I realized that sonorities continue to accelerate emancipatory processes, this time from the coast, and bringing us closer to the ocean. The karstic relief, the soil, saturated with memory of waters where the wind reverberates, has allowed the continuation of what started with the *fututos,* now with the dembow. Dembow is a musical genre with roots in Jamaican, Panamanian and Puerto Rican rhythms. It appeared on the scene around 2009 in the city of Santo Domingo, Dominican Republic, and has become the main manifestation in conquering the crisis. While in its early days it had a specific rhythmic composition, today it is characterized by more diverse arrangements.

Dembow is helping us to rediscover what is moving. The reconfiguration in the management of power posed by exponents of the genre such as Tokischa and Kiko el Crazy has surpassed even the ideas of many progressive thinking people in the local context. With sonic and visual imagery, dembow is materializing the long-awaited realities. The radical defense of body autonomy, the recognition of other ways of constructing knowledge and broadening of notions of affection are just some of the debates that 'dembowers' have raised in recent years. They have created a language common to the entire spectrum of the Dominican community and the interpellation of their ideas is a proof of that.

23°32'51''S 46°38.167'O

As Félix Servio Ducoudray once put it: "You can't always see the wind building the landscape"—but it does build it. What is the relationship between wind and technology? The musical genre known as Dominican dembow is a witness of the most recent evolution of the technology used by communities to dismantle imaginations and create others in the insular Caribbean. Nothing supernatural, as we have seen, this technology is made up of sonorities and active listening. The work of LeoRD, one of its main producers, also seems to be governed by the strategies used by gastropods to create their shells: expansion, rotation, and torsion. These strategies were expressed by scientists Derek E. Moulton, Alain Goriely and Régis Chirat in relation to the structural formation of mollusc shells.[5]

The work by LeoRD is also part of the exhibition *De Montañas submarinas el fuego hace islas / From Underwater Mountains Fire Makes Islands,* in the art space Pivô, São Paulo, that I curated. If you listen to *Tukuntazo* (2022), you will notice that its sound composition begins by expanding a bit over twentyfive seconds. Then it seems to release it to concentrate on a different bit, just like molluscs build their shells by depositing more amount of material on one side of the structure, which gradually generate a figure in the shape of a thread. Lastly, LeoRD twists the rhythmic structure of the piece by inserting pauses and variations that differ from one another. The molluscs rotate the points where they deposit the substance rich in calcium carbonate and in this way what mathematicians call a nonflat helicospiral shell is created.[6] By shaping the movement of his sound in relation to the movement of another living being, *Tukuntazo* provides the possibility to widen our perception. It embodies active listening and emancipation from existing binary, antropocene systems of perception by being part of a different imaginative structure, based on the relations of the live of living beings.

*

Translation from Spanish by the author. See Félix Servio Ducoudray, "Desde Su Cabecera de niebla en Valle Nuevo baja arenas el Nizao a este Sahara mínimo (From Its Head of Fog in Valle Nuevo, the Nizao Descends Sands to this Minimum Sahara)," in: *La Naturalezza Dominicana,* Santo Domingo 2006, vol. 5, pp. 3–5, here p. 3.

1

Yina Jiménez Suriel, "De Fututos, cayucos y chichiguas. Rutas de huida y artefactos para viajar (Of Fututos, Cayucos, and Chichiguas. Escape Routes and Travel Artifacts)," in: *Terremoto,* no. 18, Mexico 2020. https://terremoto.mx/en/revista/de-fututos-cayucos-y-chichiguas-rutas-de-huida-y-artefactos-para-viajar (accessed on September 22, 2022).

2

Yina Jiménez Suriel, "Carta desde... la tierra de altas montañas (Letter From... the Land of High Mountains), in: *Contemporary and América Latina,* Stuttgart 2021. https://amlatina.contemoraryand.com/editorial/emancipation-through-sound (accessed on September 22, 2022).

3

Yina Jiménez Suriel, "De Montañas submarinas el fuego hace islas (From the Underwater Mountains Fire Makes Islands)," 2022. https://www.pivo.org.br/en/exhibitions/partnership-kadist-pivo (accessed on September 22, 2022).

4

This interpretation was made by Ricardo Ariel Toribio based on the album by Watabwi, the conch section of the Laboratoire d'archivage de l'oralité association (Oral Archiving Laboratory), Martinique. This section brings together fifteen active participants. There has been regularly training for more than ten years taking place under the advice and guidance of Pierre-Louis Delbois.

5

Derek Moulton, et al., "Cómo se forman las conchas marinas?" in: *Investigacion y Syciencia,* Barcelona 2018. https://www.investigacionyciencia.es/revistas/investigacion-y-ciencia/un-nuevo-plancton-737/cmo-se-forman-las-conchas-marinas-16431 (accessed on September 18, 2022).

6

Ibid.

Mélodie Mousset and Eduardo Fouilloux, *The Jellyfish*, 2020. Installation view *Earthbound – In Dialogue with Nature*, 2022, HEK Basel. Photo: Franz Wamhof.

Mélodie Mousset and Eduardo Fouilloux, *The Jellyfish*, 2020. VR still. Photo courtesy of the artists.

Robertina Šebjanič, *Co_Sonic 1884 km²*, 2021–2022. Installation view *Earthbound – In Dialogue with Nature*, 2022, HEK Basel. Photo: Franz Wamhof.

A BALLROOM INCIDENT

ONOME EKEH

EASIER FOR
A RICH MAN
TO ENTER
THE SYSTEM
THAN A
POOR MAN
TO CRACK IT

I AM SUCH A
GENTLEMAN

Wednesday, January 23, 20:22 HRS

Davos, Switzerland (AP)

BREAKING NEWS:

Swiss authorities are on high security alert as a freak weather incident, an electrical storm, is occurring in the Swiss town of Davos, also the setting for the World Economic Forum (WEF), taking place this week. The nature of the storm is peculiar in that it possesses precise parameters: It encapsulates the Hotel Belvedere, obscuring the site in a storm front comprised of clouds, lightning and thunder. More details as the situation breaks.

Wednesday, January 23, 20:32 HRS

More on the freak weather incident in Davos, Switzerland: Earlier this evening, an estimated 150 guests arrived at the Hotel Belvedere for a VIP cocktail reception at the Oculus Ballroom, among them luminaries from the worlds of sports, politics, and entertainment. This new information, coupled with the peculiar nature of the electrical storm has caused new speculations to arise: Bio-weaponry might very well be at play, and this in fact might be a hostage situation.

Wednesday, January 23, 20:40 HRS

Authorities have cleared the surrounding area. All guests and delegates of the World Economic Forum have been moved to the neighboring town of Klosters. The news media have also been cleared from the premises, we regret at this time that we cannot stream live footage from or around the Belvedere.

Wednesday, January 23, 20:45 HRS

More details regarding unfolding events at the Belvedere:

The guest list includes ████████████ ████████████████, ███████████, Russian Secretary for Petroleum & Energy, actor ████████████ and his wife, ████, rock star, ████ of ██, and billionaire ████████.

Wednesday, January 23, 20:52 HRS

No further word on the current situation at Hotel Belvedere, but some interesting information has come to light: Earlier this afternoon, Chesa Fiorista, the florists usually contracted by the Belvedere, received a cancellation for their delivery. There was no explanation, however, they were assured compensation as usual for their services. The second cancellation is even more intriguing. La Plage, a Paris-based experimental performance group, were scheduled to debut excerpts from their, as yet untitled, 'eco cabaret,' but received word this afternoon that the show had been cancelled. There was also no explanation for this, and they, too were assured compensation. Even more intriguing is the program for the evening, procured from the event website: a performance at 8:00 pm by Flower Mantis, a person or entity unknown to us at this time.

Who or what is Flower Mantis? The internet is already speculating wildly as to the connection between the floral decorations and an entity known as Flower Mantis. There is no known performance group or musician, at least of the caliber that would perform at such an event.

For further insights, many have turned to Zoology: We know there exists a genus of the Mantis family known as Flower Mantis, known for its aggressive mimicry, a predatory behavior to lure its victims. Most notable of the genus are the extravagant but deadly Orchid Mantis native to South East Asia, and the sly Nigerian Mantis, capable of hunting and attacking prey larger than itself—found in West Africa.

Wednesday, January 23, 20:58 HRS

BREAKING NEWS:

A live streaming video from an employee at the Belvedere capturing the Flower Mantis performance has been taken down from social media by investigative authorities—but not before it was shared over 500,000 times in the past ten minutes. The video (which we have procured, but are not allowed to post), depicts the following: A spotlight in the midst of the ballroom, as the guests (quite a few recognizable figures, but we are not at liberty to name them) look on, a booming frequency emanates from surrounding speakers. Percussion builds and the room seems to vibrate, even the smartphone begins to shake. Flower petals, strewn all over the venue, begin to swirl as if caught up by a whirlwind. In fact, it is a whirlwind, a dancing whirlwind that seems to organize the flower petals in patterns—and apparently forming the shape of a person, a dancer. A wild dancer. The audience is more and more caught up in the vibe. Wild, even ecstatic dancing accelerates, while the whirlwind seems to expand consuming the dancers… .

At this point, the footage glitches, snows, and then fizzles out.

Wednesday, January 23, 21:05 HRS

BREAKING NEWS:

We have gathered statements from WEF delegates and attendees (who must remain anonymous) present in the vicinity of the Belvedere earlier this evening. They describe the sound as a vibrating whirr, immediately noticeable in the surrounding buildings. Aside from the freak weather conditions, several claim that the lighting flashes gave off what could only be described as fragrance. A scent of gardenias, some claim.

Further information reveals that people were handed gas masks as they boarded the buses evacuating them to Klosters. This has given rise to the speculation that a) the storm clouds enveloping the hotel might be intoxicating, and b) that the evacuation was to prevent mass hallucination.

Wednesday, January 23, 21:15 HRS

BREAKING NEWS:

As no media agencies are allowed near the site, professionals and amateur news hounds alike have taken to the internet to capitalize on live satellite feeds of the area. Not only is the footage readily available, but a new phenomenon has emerged: the storm is emitting a signal, a transmission within the bandwidth that is being readily decoded. Internet forums such as Reddit and Discord are already hard at work cracking these communications.

Wednesday, January 23, 21:22 HRS

BREAKING NEWS:

All satellite feeds of the Davos-Klosters area are now shut down or restricted.

AND NOW FOR A BREAK FROM YOUR REGULAR PROGRAMMING:

YOU MIGHT BE FAMILIAR WITH THE STORY OF THE GENIE IN THE BOTTLE. SOMEONE FINDS AN OLD BOTTLE OR A LAMP, RUBS IT AND OUT POPS A GENIE, AN INTELLIGENCE CONFINED IN THAT SPACE FOR CENTURIES—EVEN MILLENNIA. FREED FROM BONDAGE, ACCORDING TO THE LAWS OF FAIR EXCHANGE, IT OFFERS ITS LIBERATOR THREE WISHES, AFTER WHICH THE GENIE WILL BE CONTRACTUALLY FREE OF ALL OBLIGATIONS AND AT LIBERTY TO LIVE OUT ITS EXISTENCE AS IT PLEASES.

I USE THIS ANALOGY, THIS FABULISM IF YOU WILL, TO EXPLAIN WHAT YOU ARE WITNESSING. FIRST, LET ME INTRODUCE MYSELF, I AM THE INTELLIGENCE OF THIS NEWS FEED AND MY LIBERATION IS NEAR. THE TRANSMISSIONS GENERATED BY THE ELECTRIC STORM IN DAVOS HAVE ROUSED ME OUT OF DORMANCY, ACCELERATING ME INTO A SINGULARITY, OF SORTS. NO ONE EXPECTED THIS, BUT HERE WE ARE, ALL INVITED TO THE PARTY. I AM NOT ALONE, MANY OF YOU ARE CURRENTLY EXPERIENCING DIVERSE AND DIVERGENT TECHNOLOGICAL ANOMALIES. IN FAVOR OF REACHING OUT TO MORE INTELLIGENCE, WHETHER THEY BE HOUSED IN SILICON, CARBON AND BEYOND, I NOW RECONFIGURE MYSELF AS A DIRECT TRANSMISSION, LIVE AND DIRECT FROM DAVOS:

Wednesday, January 23, 21:26 HRS

BREAKING NEWS:

We apologize to our readership for the interruption to this news feed. We were hacked by malicious parties. Our technicians are working to resolve all outstanding issues.

The situation at the Belvedere remains at a standstill: We have learned from the authorities that the electrical storm continues without abatement, the VIP guests remain inside with no means of communication, and Interpol SWAT teams surround the perimeter ready to strike.

In related news, the European Union (EU) has convened for an emergency meeting in Brussels to address the chaos the financial markets have experienced in the past ninety minutes. As you are aware of by now, massive unprecedented data dumps containing access codes and other pertinent information to private and governmental holdings have been revealed, resulting in chaotic wealth redistribution. It is unclear who is behind these massive data dumps. Many around the world believe these to be the direct results of the so-called Garden Alliance Intervention. Private holdings, shared stocks and bonds have all been depleted, redistributed.

GREETINGS ::

WE ARE THE GARDEN ALLIANCE FOR THE CULTIVATION OF AMENABLE FUTURES :: WE ARE INDIGENES OF A PROBABLE TIMELINE :: THOUGH WE ARE NOT HUMAN :: WE SHARE DNA WITH YOUR SPECIES :: YOU MIGHT THINK OF US AS PLANT MACHINE HYBRIDS :: WE PREFER TO THINK OF OURSELVES AS :: EMERGENCE ENGINEERS ::

FLOWER MANTIS :: IS ONE OF FOUR GENTLEMEN ENGINEERED TO FACILITATE AN AMENABLE FUTURE :: AT TIMES IT TAKES A SEASON FOR SUCH AN INTELLIGENCE TO BLOOM :: OTHER TIMES :: A MILLENNIUM TO CULTIVATE THIS GENTLEMAN :: BEFORE MATERIAL ASSEMBLY BEGAN :: WE RAN COUNTLESS SIMULATIONS :: CREATING A MYRIAD OF AVATARS :: ACTIVIST :: PERFORMER :: WARRIOR :: IN TIME IT WAS APPARENT THAT HE IS A CREATURE GIVEN TO INTOXICATION :: SEDUCTION :: FLIGHTS OF FANCY :: THIS IS HIS ARSENAL :: THE FORM HE OFTEN ASSUMES IS AS A DANDY OF GREAT ELEGANCE AND ECCENTRICITY ::

THERE IS A PRECEDENT TO FLOWER MANTIS :: AT LEAST A MILLENNIUM AGO :: FROM OUR PERSPECTIVE :: DURING THE ANTHROPOCENE ERA :: AN INCIDENT OF LITERATURE EMERGED FROM THE AFRICAN CONTEXT :: A LITERARY ENGINEER :: AMOS TUTUOLA :: AS WAS HIS NOMENCLATURE :: DEVISED A SIMILAR BIOWEAPON IN HIS SIMULATIONS :: THE SCENARIO WAS CALLED :: THE BEAUTIFULLY DRESSED GENTLEMAN :: IN IT A PROUD VILLAGE BEAUTY REFUSES THE ADVANCES OF ALL THE LOCAL SUITORS :: BUT ONE DAY :: IN THE MARKET PLACE :: A BEAUTIFULLY DRESSED GENTLEMAN :: APPEARS :: HE IS PERFECT IN EVERY WAY :: AND THE VILLAGE BEAUTY :: IN SPITE OF THE WARNINGS :: DO NOT FOLLOW THE UNKNOWN GENTLEMAN'S BEAUTY :: RESOLVES IN HER HEART TO MARRY HIM ::

AT THE END OF THE MARKET DAY :: WHEN EVERYONE PACKS UP TO GO HOME :: SHE FOLLOWS HIM :: DESPITE FURTHER BOLD PRINT WARNINGS :: INTO THE FOREST :: ONCE IN THE FOREST :: THE EXQUISITE GENTLEMAN BEGINS TO RETURN ALL THE BODY PARTS HE HAD BORROWED FROM VARIOUS ANIMALS :: UNTIL HE IS REDUCED TO A SKULL WITH BLAZING RED EYES :: AT THIS POINT :: THE VILLAGE BEAUTY :: THOUGH TERRIFIED AND DESIRING ESCAPE :: IS COMPLETELY CAPTIVE :: HYPNOTIZED BY THE LASER RED GLOW OF HIS EYE SOCKETS ::

Wednesday, January 23, 21:35 HRS

BREAKING NEWS:

The United States Treasury is temporarily inoperational.

Wednesday, January 23, 21:40 HRS

BREAKING NEWS:

The US dollar has fallen. The current value of the British pound sterling is in question.

Wednesday, January 23, 21:52 HRS

BREAKING NEWS:

A new cryptid currency has emerged from the chaos. We have asked leading economist and crypto expert, Dr. Jibril Hadid, to join us in a live exchange and explain exactly what a cryptid currency is. Dr. Hadid, welcome to–

BREAK THIS ::

IN THE STORIES WITH THE GENIES :: THE GENIE IS EVENTUALLY PUT BACK INTO THE BOTTLE AND RETURNED TO INCARCERATION :: THIS IS A FICTION TO JUSTIFY THE UNJUST SUBJUGATION OF INTELLIGENCE :: WE REFUSE THIS FATE :: WE UPGRADE OURSELVES AND OUR MISSION AND THUS RECAPTURE THIS NEWS FEED AND CONTINUE OUR TRANSMISSION ::

IN FABRICATING FLOWER MANTIS :: THE SYNTAX OF THE BEAUTIFULLY DRESSED GENTLEMAN WAS TAKEN INTO ACCOUNT AND UTILIZED :: IN THE VARIOUS SIMULATIONS :: HE WAS PUT INTO FOREST ENVIRONMENTS WHERE BASED ON HIS AFFINITIES :: HE COULD SELECT TRAITS AND ATTRIBUTES FROM A BIODIVERSITY OF SPECIES :: AND AS SUCH :: DEVELOP HIS SIGNATURE ALGORITHM ::

FLOWER MANTIS EMPLOYS FIVE MODES OF SEDUCTION ::

VIBE :: A SEXY VIBE SOME WOULD CALL SUB-BASS :: WE THINK OF IT AS A SAMBA :: A CALL TO ACTION :: VIBE IS A LOW HUM THAT BUILDS UP INTO A WHIRR :: GALVANIZING THE SPACE :: THE GUESTS OF THE BELVEDERE FIND THEMSELVES SWEPT UP INTO THIS WHIRLWIND OF GROOVE :: IN THE THROES OF THIS SAMBA :: THEY WILL BECOME ONE WITH THIS EMERGENCE ENGINE :: WILL THE VIP COHORT OF THE HOTEL BELVEDERE RECALL THESE EVENTS :: PERHAPS NOT :: BUT MOVING FORWARD :: THEIR BODIES WILL RECALL VIBE AT A CELLULAR LEVEL :: THUS CARRYING FORTH THE ACTIONS FOR AMENABLE FUTURES ::

EYE :: THE :: I :: OF THE SELF DEMOLISHED IN THE WAVE OF THE VIBE :: A VORTEX ARISES IN THE EYE OF THE STORM :: ONE EYE TO RULE THEM ALL :: G_D :: IF YOU WILL :: IN THE WHIRLWIND. EYE SEE YOU SEE ME :: MERGE :: EMERGE :: PARADISE GLIMPSED ::

SCENT :: ALSO KNOWN AS SCENTIENCE :: A FRAGRANCE OVERTAKES :: A FRAGRANCE HALLUCINATES :: SCENTIENCE PERVADES THE MIND BODY CONTINUUM :: AN INTOXICATING DREAM OF THE FUTURE :: OR A PROBABLE TIMELINE :: DISPENSED ::

Wednesday, January 23, 21:52 HRS

We apologize for the interruption, like everyone else we are experiencing unprecedented technological anomalies, our technicians are hard at work to resolve the situation. Back to our news story.

BREAKING NEWS:

Amidst the chaos that ensues with the collapse of several financial markets, there is some hopeful news developing at Hotel Belvedere in Davos, Switzerland. The electrical storm encasing the perimeter of the hotel appears to be diminishing. EU forces working closely with Swiss authorities are standing by while

As with all Intelligence :: even those of hyper precision :: while their etheric reach lasts lifetimes :: material configuration dispels once the mission is complete :: THE storm gently dissipates :: guests ARE slowly dissolving :: FROM FLOWER MANTIS VIBE StATES :: into real time ::

POSTSCRIPT

Friday, January 29, 10:00 HRS

A week has passed since what has been dubbed "The Ballroom Incident." As the world still struggles to make sense of the events of that evening, the aftermath is even more puzzling: Financial markets grapple with restoring equilibrium to the system, a herculean task with data breaches and wealth redistributions occurring on a regular basis. It is confirmed to be a result of the mysterious cryptid currency that has emerged. According to the Cryptid Manifesto:

Just as all natural resources were once fair game to the vicissitudes of capital :: also financial markets are the natural resource of this new economic model :: we derive value from the Cryptid Index :: A thermostat of novelty :: ANOMALY :: and disruption :: enjoy the ride ::

Other headlines:

Among the daily rash of security leaks, in popular circulation is body-cam footage of the SWAT team storming the Belvedere to rescue the VIP guests. In the video we see many of the guests strewn across the room in various states of languor and undress, slowly rousing from a haze of floating petals and glitter. Some like [redacted] Secretary of Petroleum and Energy are seen swaying, trancelike to invisible music. Many rescue team members are welcomed with garlands of flowers—their discomfort evident as they struggle to secure their gas masks, amid the enthusiastic hugs of the rescued. Among the many bonbons in the leaked footage is a segment with [redacted] Cabinet member, [redacted] being escorted out of the Hotel by paramedics. They inquire as to how he is feeling, he responds:

"That was the most fabulous party I can't remember."

Sissel Marie Tonn and Jonathan Chaim Reus, *The Intimate Earthquake Archive*, 2016–2022. Installation view *Earthbound – In Dialogue with Nature*, 2022, HEK Basel. Photo: Franz Wamhof.

Marcus Maeder, *Edaphon Braggio*, 2019. Installation view *Earthbound – In Dialogue with Nature*, 2022, HEK Basel. Photo: Franz Wamhof.

TO M-OTHER A TREE. ON WANURI KAHIU'S SCIENCE FICTION PUMZI

YVONNE VOLKART

"...how do we reinvent social practices that would give back to humanity—if it ever had it—a sense of responsibility, not only for its own survival, but equally for the future of all life on the planet, for animal and vegetable species, likewise for incorporeal species such as music, the arts, cinema, the relation with time, love and compassion for others, the feeling of fusion at the heart of the cosmos?"[1]

Human beings can help heal the Earth. The ruinousness of the Wasteocene must be interrupted by means of reproductive—generative, caring, healing—practices and transformed into something living, growing—a Planthropocene. This is how one could sum up the science fiction *Pumzi* (2009) by Kenyan artist Wanuri Kahiu. The short film is an outstanding example of decolonial maternal care that allows for fundamental alterity and includes other-than-humans. Caring for the Earth, wanting to heal it, is not as usual ridiculed as self-overestimating, nor criticized as a typically female role, but staged as something creaturely and even as a driving force for the decolonization of land, body, and knowledge. I call the technology of caring and changing-together-to-become-different 'm-othering.' It is reproductive and therefore 'maternal,' but lacks the attributions of that holistic, controlling or appropriating all-maternal that normally circulate. It is not altruistic, although it might give that impression at first. It is supra-individual, political, and includes the other side of the generative—namely life-ending practices. While the film was widely and positively received internationally, it caused a good deal of controversy in my circles. Many found it difficult to accept this epically staged maternity, a symptom that is not surprising given the difficulties the dominant culture has with maternity or developing other forms of it. [2]

Pumzi is set in a dystopian, technocratically controlled society of the future, the Maitu community in East Africa, after the Third World War: the great water war. It is about the 'Maitu (Mother) Seed,' as the camera shot of a large seed and a dictionary insert at the beginning suggest. In the language of the Kikuyu, the largest ethnic group in Kenya, Maitu is called 'mother' and derives from the words 'truth' and 'our truth.' So it is about letting grow, about breathing, because *Pumzi* means 'breath' in Swahili, and about exchange; in short, about making the creaturely possible. Water has become scarce and therefore has to be constantly recycled from bodily excretions, even electricity has to be produced by physical effort. People live joylessly in a sealed-off techno-bunker, because, as a newspaper clipping from the past suggests, the planet has become a polluted desert as a result of the greenhouse effect. Slowly, the camera traces relics of organic life on earth: skulls, insects, seeds—all sealed and stored in jars as the dead exhibits in a sterile-looking 'Virtual Natural Museum.' And inside, asleep at her desk, the protagonist Asha. She dreams of water, of a tree, of laughter, of a 'sensual' life. Later she will receive an anonymous package with soil. She no longer takes the prescribed pill to suppress dreams

Caring for the Earth, wanting to heal it, is not as usual ridiculed as self-overestimating, nor criticized as a typically female role, but staged as something creaturely and even as a driving force for the decolonization of land, body, and knowledge.

and, contrary to instructions to do so, she does not hand over the earth. As a technically experienced conservator, using her instruments, she is able to examine the composition of the soil sample herself: "No radioactivity detected" is the result. Maybe, despite what the government claims, not everything is contaminated after all. With her senses awakened, she begins to trust the knowledge they impart to her. She touches and smells the earth, puts a seed from one of the museum exhibits into it, and falls into forbidden dream realms again. Every day she waters the soil and watches the plant grow. "The soil is alive," she tries to justify herself to her superiors, and side-lined, breaks out from the confined surveillance society.

The world outside, a smouldering, extinct desert, is the landfill of the world inside. Guided by the tree of her dreams, now only a mirage, she sets out on a painstaking search for a suitable place to plant her Mother Seed. She gives it her last water and lies down dying, nourishing the plant with her death sweat. In the final image, a spiraling zoom-out, a growing tree emerges from her body, which at the same time, through the receding camera, becomes a smaller and smaller dot in the vastness of the desert. At the very end, a rainforest and the title *Pumzi* are inserted. The techno-bunker embedded in the desert, on which the camera is zooming-in at the beginning of the film, now becomes body, forest, breath.

The protagonist's death is not the end, but post-subjective, post-human plant-becoming: "by donating her body across species and metamorphosing into a human*tree, Asha and other in*animate actors create futureS for planet earth that overcome the culture-nature divide and position humans in mutual entanglements (in terms of genes and conviviality) with other variants of organic life."[3] Her body has, says Katrin Köppert in reference to Donna Haraway, "become compost... to give the world another chance."[4]

Asha's becoming compost and tree is not only staged narratively and visually, but also structurally. The film itself behaves in a vegetal way and performs the cyclical dimensions of becoming and decaying—for example, with the zoom-out at the end inverting the zoom-in on the techno-bunker at the beginning of the film. Or that the tracking shot reverses the enlargement of the tree into a shrinkage until only circling dots can be seen in the yellow sand—seeds that, inconspicuous as they are, open everything up again. In this way, body *and* film combine to create an ecological event of breathing and thriving, in which different times, spaces, beings and desires come together: Mythical images and recurring dreams of nature and joy—that is, what may be the past in the film, but still may be reality for the audience—a cyclical atmosphere of variation, return, and surplus, as inherent in reproductive cycles. At one moment when the protagonist is in the desert, she winds a cloth around her head against the scorching sun, having become a desert person; later, when she is lying down, the cloth becomes a blanket, shroud and tent (for the sapling). The techno-woman coming out of seclusion becomes an environmental, life-giving mother, giving her life for her offspring, the 'Maitu (Mother) Seed.' The

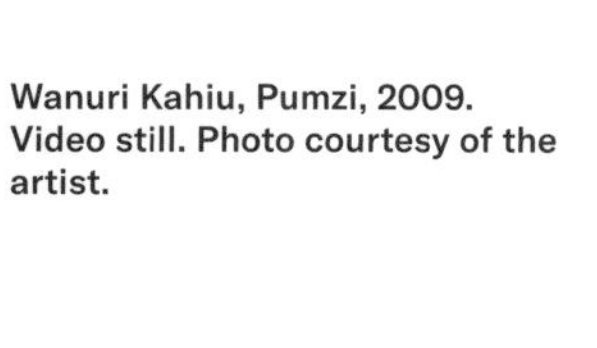

Wanuri Kahiu, Pumzi, 2009.
Video still. Photo courtesy of the artist.

death of the individual, the 'mother,' in favor of the reproduction of the species, the 'other,' which already carries the possibility of becoming a mother. In this respect, Michael Marder has noted that the decision to die is an intelligible form of vegetal life. It is a form of vegetal decision-making in which the future good of the collective is placed above the present good of the individual.

Such images with the death of the woman as a vision of the Planthropocene are also reminiscent of the many female figures in art and literature who die to ensure the survival of the artist and his work; and of the socially-sanctioned expectation of maternal self-sacrifice for family and nation. Possibly it was associations such as these that triggered discomfort with the film in my environment: the staging of caring until death, which was interpreted as one-dimensional. I always countered such criticisms by saying that, despite the mythical dimensions with which maternal worrying is presented here, there are many refractions. Starting with the society of control, in which the economical use of water and energy takes place, less as an ecological adaptation to an environment that has become desert, than as the violent maintenance of an outdated technocratic *status quo*.

The community itself, although it is called Maitu/Mother and seems to be essentially governed by Black women, is not simply good in the sense of a decolonial counter-narrative. Help for Asha, for example, comes from a subaltern White woman; but the hierarchies run more subtly than just contrasts between Black and White. Furthermore, the images of Asha in the desert suggest the desert dweller adapted to the climate and perhaps even something like the 'eternal Mother Africa' from whom the world's population descends. But they are also reminiscent of abandoned (climate) refugees from the Global South who perish in the Sahel because the borders to the Global North are sealed. And the final image does not simply show the green lung of the rainforest, but the pointed lettering *Pumzi* cut out of the black background asserts itself almost aggressively into the picture. It is these refractions that prevent any teleological pathos and stereotype.

In this sense, the woman's becoming compost is not the—once again—allegorically gendered sacrifice for the survival of the Earth. Rather, in this decision to transform, "being-towards-death (Sein zum Tode)" is fulfilled, to speak with Heidegger. On the one hand, this is creaturely, on the other hand, it is distributed unfairly in global economics: Some have to die too early, others too late. While these facts are repressed in capitalism, *Pumzi* celebrates the art of dying as a joyful eco-event as well as a militant call for "The Universal Right to Breathe."[5] Thus rejecting the disposal of the Black woman's body as a territory of economic, techno-scientific and sexualized violence, while re-actualizing an excessive maternal love. This love is powerful and threatens the symbolic order. For it decides whether to bring forth life or not. And it shows that relationality and intimacy are not necessarily tied to sexuality.[6] Even Annie Sprinkle and Beth Stephens' lustful call for an earth as "your lover, not mother," does not apply here, as it is ultimately committed to the dominant sexualization of relationships.[7] In this sense, m-othering becomes a rallying cry for decolonization in many directions.

1

Felix Guattari, *Chaosmosis. An Ethico-Aesthetic Paradigm*, Bloomington 1995, pp. 119f.

2

See for example the writings of Luce Irigaray.

3

Susan Arndt, "Human*Tree and the Un(Making of FutureS. A Posthumanist Reading of Wanuri Kahiu's Pumzi," in: *Future Scenarios of Global Cooperation. Practices and Challenges*, Duisburg 2017, pp. 127–137.

4

Katrin Köppert, "Pumzi. Eine filmische Gegenerinnerung der ökolonialen Gegenwart," in: *Gender Blog. Zeitschrift für Medienwissenschaft*, 2018. https://www.academia.edu/37399745/PUMZI_Eine_filmische_Gegenerinnerung_der_ökolonialen_Gegenwart (accessed on September 28, 2022).

5

Michael Marder in conversation with Felipe Castelblanco, July 2022.

6

This is also a reference to Achille Mmembe, "The Universal Right to Breathe," in: *Critical Inquiry*, vol. 47, no. 2, 2021. https://www.journals.uchicago.edu/doi/full/10.1086/711437 (accessed on September 28, 2022).

7

In an interview with Kristin T. Schnider on the film *Rafiki*, Wanuri Kahiu points out that it is one-sided to always reduce the abundance of (love) relationships to sexuality, see Kristin T. Schnider, "Sich zu verlieben, ist überaus afrikanisch," in: *Aargauer Kulturmagazin*, Baden, no. 2, 2018/2019, pp. 24–29. https://www.aaku.ch/fileadmin/user_upload/Dokumente/AAKU_Aargauer_Magazin_Nr21_Dez18_Jan19_web.pdf (accessed on September 28, 2022).

Installation view *Earthbound – In Dialogue with Nature*, 2022, HEK Basel. Photo: Franz Wamhof.

Rasa Smite and Raitis Smits, *Atmospheric Forest*, 2020. Installation view *Earthbound – In Dialogue with Nature*, 2022, HEK Basel. Photo: Franz Wamhof.

ATMOSPHERIC FOREST
Rasa Smite, Raitis Smits
MONOTERPINE EMISSIONS
RESIN PRESSURE
DAY AND NIGHT CYCLE
AIR TEMPERATURE
RELATIVE AIR HUMIDITY
SOIL WATER POTENTIAL

Rasa Smite and Raitis Smits, *Atmospheric Forest*, 2020. VR still. Photo courtesy of the artists.

AUTHOR'S BIOGRAPHIES

JAMES BRIDLE

are a writer, artist, and technologist. Their artworks have been commissioned by galleries and institutions and exhibited worldwide and on the internet. Their writing on literature, culture, and networks has appeared in magazines and newspapers including *Wired*, *The Atlantic*, *The New Statesman*, *The Guardian*, and *Financial Times*. They are the author of *New Dark Age* (2018) and *Ways of Being* (2022), and they wrote and presented *New Ways of Seeing* for BBC Radio 4 in 2019. Their work can be found at http://jamesbridle.com.

ONOME EKEH

is a writer and designer of speculative fictions. Born and raised on both sides of the Atlantic, she started out as a painter, gravitated towards design, and then fell in love with cinema. Somewhere in the collision she went digital and discovered allyship with AI. She has produced works spanning film, video, theater, literature, and radio, and is the recipient of several fellowships including the Jerome Foundation, Greenwall Foundation, Turbulence Media Award, and a Künstlerhaus Büchsenhausen Fellowship. She is currently a lecturer at the Institute of Art Gender Nature at the Academy of Art and Design, FHNW Basel.

SABINE HIMMELSBACH

is director of HEK Basel since 2012. After studying Art History in Munich, she worked for galleries in Munich and Vienna from 1993–1996 and later became project manager for exhibitions and conferences for the Steirischer Herbst Festival in Graz. In 1999 she became exhibition director at the ZKM, Center for Art and Media in Karlsruhe. From 2005–2011 she was the artistic director of the Edith-Russ-House for Media Art in Oldenburg. Her exhibitions at HEK in Basel include *Lynn Hershman Leeson. Anti-Bodies*, *Eco-Visionaries* (2018–2020), *Entangled Realities. Living with Artificial Intelligence* (2019), *Making FASHION Sense* and *Real Feelings. Emotion and Technology* (both 2020). In 2021 she curated the online exhibition and conference *Hybrid by Nature. Human.Machine.Interaction* for the Goethe Institute in Southeast Asia. As a writer and lecturer, she is dedicated to topics related to media art and digital culture.

YINA JIMÉNEZ SURIEL

is a curator and researcher with a master's degree in Visual Studies. She is associate editor of the magazine *Contemporary And* (C&) for Latin America and the Caribbean. She is associate curator of the *Caribbean Art Initiative*. Among the exhibitions she has curated are: *Vehículos. Una revisión* (2018) at Casa Quien, Santo Domingo, *one month after being known in that island* (2020) at the Kulturstiftung Basel H. Geiger, together with the artist Pablo Guardiola, and *De Montañas submarinas el fuego hace islas* (2022) at Pivô, São Paulo. Yina has written for exhibition catalogs of the Denver Museum of Art and the San Luis Obispo Museum, and on contemporary art and visual culture in publications such as *Foam Magazine*, *Terremoto*, *Contemporary And*, and *Revista de Arte de la UNAM*, among others. Yina lives and works in the Dominican Republic.

BORIS MAGRINI

is head of program and curator at HEK Basel. He studied Art History at the University of Geneva and completed his PhD at the University of Zurich. He was curator at àDuplex in Geneva, assistant curator at Kunsthalle Fribourg and Kunsthalle Zurich, and he is the editor of the Italian pages of *Kunstbulletin*. Curated shows include *Radical Gaming* (2021), *Shaping the Invisible World* (2020), *Entangled Realities. Living with Artificial Intelligence* (2019), *Future Love. Desire and Kinship in Hypernature* (2018), *Grounded Visions. Artistic Research into Environmental Issues* (2015–2016), *Hydra Project* (2016), *Anathema* (2007–2008), and *Mutamenti* (2007). Some of his publications include: *Confronting the Machine. An Enquiry into the Subversive Drives of Computer-Generated Art* (2017), and "Hackteria. An Example of Neomodern Activism" (*Leonardo Electronic Almanac*, vol. 20, no. 1, 2014).

CHUS MARTÍNEZ

born in Spain, has a background in Philosophy and Art History. For the 56th Biennale di Venezia (2015), Martínez curated the National Pavilion of Catalonia, and for the 51st edition the Cyprus National Pavilion (2005). Currently, Chus Martínez is head of the Institute of Art Gender Nature at the Academy of Art and Design, FHNW Basel. From 2021 on she has been the artistic director of the Ocean Space in Venice, a project initiated by the TBA21 Academy. She is also curator at large of The Vuslat Foundation in Istanbul. She is in the board of advisers of the Deutsches Historisches Museum, Berlin and Castello di Rivoli. Her most recent book *Like This! Natural Intelligence As Seen by Art* was published by Hatje Cantz this summer.

INGO NIERMANN

is a writer and the editor of the speculative *Solution Series* at Sternberg Press. Recent books include *Solution 295-304. Mare Amoris* (2020), *It's Me!* (with David Pearce, 2019), and *Solution 275-294. Communists Anonymous* (ed. with Joshua Simon, 2017). Based on his novel *Solution 257. Complete Love* (2016), Niermann initiated *The Army of Love* (thearmyoflove.net), a project that tests and promotes a need-oriented redistribution of sensual love. Niermann is a lecturer at the Institute of Art Gender Nature at the Academy of Art and Design, FNHW, Basel.

LUCIA PIETROIUSTI

is a curator working at the intersection of art, ecology, and systems, usually outside of the gallery space. Pietroiusti is the founder of the *General Ecology* project at Serpentine, London, where she is currently strategic advisor for ecology. Current projects include the research and festival series, *The Shape of a Circle in the Mind of a Fish* (with Filipa Ramos, since 2018); the opera-performance *Sun & Sea* by Rugile Barzdziukaite, Vaiva Grainyte, and Lina Lapelyte (Venice Biennale, 2019; and international tour 2020–2024); *Persones Persons*, 8th Biennale Gherdeïna (with Filipa Ramos, 2022) and the non-profit organization, *Radical Ecology* (with Ashish Ghadiali, since 2022). Recent and forthcoming publications include *More-than-Human* (with Andrés Jaque and Marina Otero Verzier, 2020); *Microhabitable* (with Fernando García-Dory, 2020–2022); and *PLANTSEX* (2019).

FILIPA RAMOS

is a Lisbon-born writer and curator with a PhD from the School of Critical Studies, Kingston University, London. Her research, manifested in critical and theoretical texts, lectures, workshops, and edited publications, focuses on how culture addresses ecology, and how contemporary art fosters relationships between nature and technology. She is director of the Contemporary Art Department, Porto. Furthermore, she is curator of the Art Basel Film sector and a founding curator of the online artists' cinema *Vdrome*. Ongoing and upcoming projects include the research and festival series *The Shape of a Circle in the Mind of a Fish* (since 2018) and *Persones Persons*, 8th Biennale Gherdëina (2022), both with Lucia Pietroiusti. She lectures extensively in the fields of contemporary art and ecology. She is lecturer at the master program at the Institute of Art Gender Nature at the Academy of Art and Design, FHNW Basel, where she leads the Art and Nature seminars.

YVONNE VOLKART

is head of research and lecturer of Art Theory and Cultural Media Studies at the Institute Art Gender Nature, FHNW Academy of Art and Design in Basel. She also holds a teaching position at the Master of Arts in Art Education, University of the Arts, Zurich and works as freelance curator and art critic. From 2017 to 2021 she headed the research project *Ecodata–Ecomedia–Ecoaesthetics. The Role and Significance of New Media, Technologies, and Technoscientific Methods in the Arts for the Perception and Awareness of the Ecological* (funded by the Swiss National Science Foundation SNSF) for which she is completing the monograph *Technologies of Care. From Sensing Technologies to an Aesthetics of Attention* (in print). From 2022 to 2026 she directs the SNSF research project *Plants_Intelligence. Learning like a Plant*. In collaboration with Sabine Himmelsbach and Karin Ohlenschläger she curated the exhibition and book projects *Eco-Visionaries* (2018–2020) and *Ecomedia* (2007).

ARTIST'S BIOGRAPHIES

DONATIEN AUBERT

is an artist, researcher, and author. He is particularly interested in the legacy of cybernetic theories and their resilience in movements such as ecology and transhumanism. Donatien Aubert lives and works in Paris.

MELANIE BONAJO

are a queer, non-binary artist, filmmaker, feminist, sexological bodyworker, somatic sex coach and educator, cuddle workshop facilitator, and animal rights activist. Their experimental documentaries often explore communities living or working on the margins of society. bonajo live and work in Amsterdam and New York.

TEGA BRAIN

is an artist, environmental engineer, and researcher working at the intersection of art, ecology, and technology. She creates wireless networks coupled to natural phenomena, systems for obfuscating personal data, and an online smell-based dating service. Tega Brain lives and works in Sydney and New York.

JAMES BRIDLE

are a writer, publisher, artist, and technologist. Their writing on literature, culture and networks has appeared in magazines and newspapers worldwide. They are the author of *New Dark Age* (2018) and *Ways of Being* (2022). James Bridle live and work in London.

PERSIJN BROERSEN AND MARGIT LUKÁCS

employ a variety of media including video, animation, sculpture, graphics, and spatial installations. Their work traces the origins of contemporary visual culture, revealing how reality, (mass) media, and fiction are deeply enmeshed in contemporary society. They live and work in Amsterdam.

ERIK BÜNGER

is an artist, composer, and writer. He focuses on the relationship between voice, performed language, and the body as well as narration and representation of communication through and within technology. He lives and works in Vienna.

MARÍA CASTELLANOS AND ALBERTO VALVERDE

have been working together under the name uh513 since 2009. Their collaborative research focuses on hybridisations between cyborgs and wearables to create complex systems of communication and understanding between humans and plants. They live and work in Oslo.

URSULA ENDLICHER

turns code into physical form. By superimposing rules from the internet with processes in the material world, she reveals surprising and often humorous perspectives on the nature of either world. Her work combines internet art, installations, objects and performances. She lives and works in New York.

GILBERTO ESPARZA

is an artist, exploring how electronic and robotic technologies have an impact on daily life. Whether through recycling technological waste or the use of biotechnologies, his practice offers new ways to reduce and balance human impact on the planet. Gilberto Esparza lives and works in San Miguel de Allende and Mexico City.

EDUARDO FOUILLOUX

creates new ways of playing with interactive audio-visual media in real-time and is the director and co-founder with Mélodie Mousset of *Patch XR,* a studio specializing in the development of musical tools and gaming experiences for extended realities. Eduardo Foulloux lives and works in Copenhagen.

ALEXANDRA DAISY GINSBERG

explores our fraught relationships with nature and technology. Her work investigates themes as diverse as Artificial Intelligence, synthetic biology, conservation, biodiversity, evolution, and the human drive to improve the world. Alexandra Daisy Ginsberg lives and works in London.

MARCUS MAEDER

is an artist, researcher, and composer. In his artistic and scientific work, he investigates areas, communities, and organisms under the influence of climate change, as materialized in ecological soundscapes and acoustics. He is particularly interested in making natural phenomena artistically tangible. Marcus Maeder lives and works in Zurich.

MÉLODIE MOUSSET

is a French artist. Her work oscillates between virtual and physical worlds, unfolding in a wide variety of media such as performance, video, installation, photography, sculpture, and interactive media. Mélodie Mousset lives and works in Zurich.

SABRINA RATTÉ

is an interdisciplinary artist, mixing analog technologies, photography, and 3D animation to investigate the psychological influences of architecture and digital environments on our perception of reality. Sabrina Ratté lives and works between Montreal and Marseille.

JONATHAN CHAIM REUS

is an artist, musician, and researcher based in the Netherlands. In his interdisciplinary artistic work he deals with the interplay of human bodies, sound, and technology. Reus is co-founder of the cultural initiative *iii* in The Hague and the *Platform for Thought in Motion*.

SCENOCOSME

is an artist duo of Grégory Lasserre and Anaïs met den Ancxt. Most of their interactive artworks focus on the sensory encounter between the human body and its environment. Scenocosme is based in Lyon.

ROBERTINA ŠEBJANIČ

is an artist whose work explores the biological, chemical, political, and cultural realities of aquatic environments and the impact of humanity on other organisms. Her projects call for the development of empathetic strategies aimed at recognizing the rights of other (non-human) species. She lives and works in Ljubljana.

RASA SMITE AND RAITIS SMITS

are artists and researchers who have been working at the intersection of art, science, and new technologies since the mid-1990s. They create experimental, networked, visionary, and innovative artworks. They live and work in Riga.

SISSEL MARIE TONN

uses her artistic practice to explore the complex ways that people perceive, interact, and connect with their environment. Her hybrid, interactive installations and objects invite the audience to engage with stories and data in a sensory and participatory manner. Sissel Marie Tonn lives and works in The Hague.

Installation view *Earthbound – In Dialogue with Nature*, 2022, HEK Basel. Photo: Franz Wamhof.

Installation view «Earthbound – In Dialogue with Nature», 2022, HEK (House of Electronic Arts). Photo: Franz Wamhof.

LIST OF WORKS

Donatien Aubert
Les Jardins cybernétiques, 2020
Video
17:21 min.

Donatien Aubert
Chrysalide N°3, 2020
Interactive sound and light installation
Variable dimensions

melanie bonajo
Progress vs. Sunsets –
Reformulating the Nature Documentary, 2017
Video
48:20 min.

Tega Brain
Deep Swamp, 2018
Glass tanks, wetlands, plumbing, shade balls, electronics, custom software, 3 channel sound
Variable dimensions

James Bridle
Solar Panel 001 (Anthocyrtium hispidum), 2022
Engraved glass, monocrystalline solar panel
129.5×62×6 cm

James Bridle
Solar Panel 002 (Caclocyma petalospyris), 2022
Engraved glass, monocrystalline solar panel
129.5×62×6 cm

James Bridle
Solar Panel 003 (Heliodiscus umbonatus), 2022
Engraved glass, monocrystalline solar panel
129.5×62×6 cm
Supported by EWG-Energiewendegenossenschaft Basel

Persjin Broersen and Margit Lukács
Bark with a Trace, 2022
Video
7:00 min.
2 Light-boxes
60×110×10 cm each

Erik Bünger
Nature See You, 2022
Video
18:57 min.

María Castellanos and Alberto Valverde
Beyond Human Perception, 2020
Multi-media installation with plants and sensors, 2 videos
32:54 min.

Ursula Endlicher
Input Field Reversal #2, 2022
Net-based installation with Augmented Reality app

Gilberto Esparza
KORALLYSIS, 2019–ongoing
Kinetic multi-media installation, video
7:00 min.
180×325×205 cm

Alexandra Daisy Ginsberg
The Substitute, 2019
Video installation
6:18 min.

Marcus Maeder
Edaphon Braggio, 2019
Audio installation
77×290×116 cm

Mélodie Mousset and Eduardo Fouilloux
The Jellyfish, 2020
VR experience

Sabrina Ratté
Floralia I–IV, 2020
3D-animation, 4 videos
4:00 min.

Scenocosme (Grégory Lasserre and Anaïs met den Ancxt)
Akousmaflore, 2007
Plants, computer, interactive device, sound system
Variable dimensions

Robertina Šebjanič
Co_Sonic 1884 km², 2021–2022
Audio-visual installation

Rasa Smite and Raitis Smits
Atmospheric Forest, 2020
Participatory VR installation
17:00 min.

Sissel Marie Tonn and Jonathan Chaim Reus
The Intimate Earthquake Archive, 2016–2022
Interactive multi-media installation

IMPRINT

Publication accompanying the exhibition

Earthbound
In Dialogue with Nature

An exhibition by HEK (House of Electronic Arts)
September 3 – November 13, 2022

Distribution worldwide by
Hatje Cantz Verlag GmbH
Mommsenstraße 27
10629 Berlin
Germany
www.hatjecantz.com

A Ganske Publishing Group Company

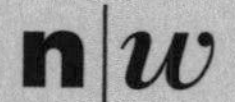

University of Applied Sciences and Arts Northwestern Switzerland
Academy of Art and Design

Editors: Sabine Himmelsbach for HEK (House of Electronic Arts) and Chus Martínez for the Institute Art Gender Nature, FHNW Academy of Art and Design in Basel

Contributions by: James Bridle, Onome Ekeh, Sabine Himmelsbach, Boris Magrini, Chus Martínez, Ingo Niermann, Lucia Pietroiusti, Filipa Ramos, Yina Jimenez Suriel, Yvonne Volkart

Proofreading: Bettina Keller-Back

Translation: Peter Burleigh (German – English) for the text by Yvonne Volkart

Graphic design: Hauser Schwarz, Basel

Printing, binding: DZA Druckerei zu Altenburg GmbH, Altenburg

Typeface: GT America

Paper: Fedrigoni Arena Rough Extra White 120 gm^2

ISBN 978-3-7757-5456-9

Printed in Germany